Verena Raschke-Cheema

Health Benefits of Traditional East African Foods and Food Habits

Verena Raschke-Cheema

Health Benefits of Traditional East African Foods and Food Habits

A detailed investigation

Südwestdeutscher Verlag für Hochschulschriften

Impressum / Imprint
Bibliografische Information der Deutschen Nationalbibliothek: Die Deutsche Nationalbibliothek verzeichnet diese Publikation in der Deutschen Nationalbibliografie; detaillierte bibliografische Daten sind im Internet über http://dnb.d-nb.de abrufbar.
Alle in diesem Buch genannten Marken und Produktnamen unterliegen warenzeichen-, marken- oder patentrechtlichem Schutz bzw. sind Warenzeichen oder eingetragene Warenzeichen der jeweiligen Inhaber. Die Wiedergabe von Marken, Produktnamen, Gebrauchsnamen, Handelsnamen, Warenbezeichnungen u.s.w. in diesem Werk berechtigt auch ohne besondere Kennzeichnung nicht zu der Annahme, dass solche Namen im Sinne der Warenzeichen- und Markenschutzgesetzgebung als frei zu betrachten wären und daher von jedermann benutzt werden dürften.

Bibliographic information published by the Deutsche Nationalbibliothek: The Deutsche Nationalbibliothek lists this publication in the Deutsche Nationalbibliografie; detailed bibliographic data are available in the Internet at http://dnb.d-nb.de.
Any brand names and product names mentioned in this book are subject to trademark, brand or patent protection and are trademarks or registered trademarks of their respective holders. The use of brand names, product names, common names, trade names, product descriptions etc. even without a particular marking in this work is in no way to be construed to mean that such names may be regarded as unrestricted in respect of trademark and brand protection legislation and could thus be used by anyone.

Coverbild / Cover image: www.ingimage.com

Verlag / Publisher:
Südwestdeutscher Verlag für Hochschulschriften
ist ein Imprint der / is a trademark of
OmniScriptum GmbH & Co. KG
Heinrich-Böcking-Str. 6-8, 66121 Saarbrücken, Deutschland / Germany
Email: info@svh-verlag.de

Herstellung: siehe letzte Seite /
Printed at: see last page
ISBN: 978-3-8381-5079-6

Zugl. / Approved by: Vienna, Austria, Univeristy of Vienna, Diss., 2007

Copyright © 2015 OmniScriptum GmbH & Co. KG
Alle Rechte vorbehalten. / All rights reserved. Saarbrücken 2015

TABLE OF CONTENTS

Page

LIST OF TABLES 5
LIST OF FIGURES 6
LIST OF ABBREVIATIONS 7
INTRODUCTION AND OBJECTIVES 9
DEFINITIONS 12

1. NEED FOR AN ONLINE COLLECTION FOR TRADITIONAL AFRICAN FOOD HABITS 14

    1.1. Background and objectives for the rational to develop an online collection of traditional African food habits 14
    1.2. Methods to investigate the need for an online collection of traditional African food habits 18
        1.2.1. Systematic review of online collections 18
        1.2.2. Questionnaire administration at the 18th ICN 19
    1.3. Results 20
        1.3.1. Systematic review of online collections on African food habits 20
        1.3.2. Questionnaire responses from opinion leaders in the nutritional sciences 22
    1.4. Discussion 25
    1.5. Résumé 28
    1.6. References 30
    1.7. Appendix A 36

2. CONTENT OF AN INNOVATIVE ONLINE COLLECTION OF TRADITIONAL EAST AFRICAN FOOD HABITS (1930s-1960s): DATA COLLECTED BY THE *MAX -PLANCK NUTRITION RESEARCH UNIT*, BUMBULI, TANZANIA 40

    2.1. Background and objectives for the development of an online collection of traditional East African food habits 40
    2.2. Methods used for the development of the online collection 43
        2.2.1. The *Oltersdorf Collection* 43
        2.2.2. Data extraction and classification 44
        2.2.3. The online collection 44
    2.3. Content of the online collection 46
    2.4. Discussion 57
    2.5. Résumé 61
    2.6. References 62

3. **INVESTIGATION OF DIETARY INTAKE AND HEALTH STATUS IN EAST AFRICA IN THE 1960s: A SYSTEMATIC REVIEW OF THE HISTORIC *OLTERSDORF COLLECTION*** 74

    3.1. Background and objectives for the systematic review of the *Oltersdorf Collection* 74
    3.2. Methods of the systematic review 77
        3.2.1. The *Oltersdorf Collection* 77
        3.2.2. Data extraction and classification 77
    3.3. Results of the systematic review 78
        3.3.1. Systematic review process 78
        3.3.2. Limitations of the research 80
        3.3.3. Funding sources 80
        3.3.4. Overview of the research 81
        3.3.5. Outcomes related to dietary intake and adequacy 83
        3.3.6. Outcomes related to health status 87
    3.4. Discussion 91
        3.4.1. Dietary intake and adequacy 91
        3.4.2. Health status 94
        3.4.3. Relationships between dietary intake/adequacy and health status 96
    3.5. Résumé 97
    3.6. References 107

4. **COLONIAL AND NEOCOLONIAL FORCES AND THE ERADICATION OF TRADITIONAL FOOD HABITS IN EAST AFRICA: HISTORICAL PERSPECTIVE ON THE *NUTRITION TRANSITION*** 124

    4.1. Colonial impact on the *nutrition transition* in East Africa 128
        4.1.1. Development of international trade routes 129
        4.1.2. Seizure of arable land 131
        4.1.3. Creation of cash crop economies 131
        4.1.4. Ecological destruction and the loss of hunter-gatherer areas 132
        4.1.5. Introduction of agricultural techniques and ecological degradation 132
        4.1.6. Displacement of indigenous crops 133
        4.1.7. Loss of indigenous markets 134
        4.1.8. Cultural indoctrination (i.e. "Education") 134
    4.2. Neocolonial impact on the *nutrition transition* in East Africa 135
        4.2.1. Suppression of domestic self-sustainability 137
        4.2.2. Dependence on low-quality staple food imports 137
        4.2.3. Continued displacement of indigenous crops 139
        4.2.4. Increased availability of processed food products 142
        4.2.5. The rise of 'super'markets 143
        4.2.6. Consequences of urbanization 144
        4.2.7. Low-quality non-home prepared foods 145
        4.2.8. Disruption of the family unit 146
        4.2.9. Disparities of socio-economic status 147
        4.2.10. Nutrition-related propaganda (i.e. advertising) 148

4.2.11. Engineering of famines 149
4.2.12. Eradication of famine foods 150
4.2.13. Consequences of disease epidemics 152
4.3. Résumé **153**
4.4. References **154**

5. IS THE NON-COMUNICABLE DISEASE EPIDEMIC IN EAST AFRICA BEING CAUSED BY A LOSS OF TRADITIONAL FOOD HABITS? 175

5.1. Objectives for the systematic review **175**
5.2. Methods of the systematic review **177**
   5.2.1. Criteria for considering studies 177
   5.2.2. Search protocol 178
5.3. Results of the systematic review **179**
   5.3.1. Studies included and excluded 179
   5.3.2. Overview of subjects 180
   5.3.3. Overview and categorization of cohorts 181
   5.3.4. Methodologies 184
   5.3.5. Outcomes 185
5.4. Discussion **193**
5.5. Résumé **197**
5.6. References **212**

6. UTILIZATION AND FUTURE APPLICATION FOR AN ONLINE COLLECTION FOR TRADITIONAL AFRICAN FOOD HABITS 220

6.1. Background and objectives for the web site evaluation **220**
6.2. Methods **222**
   6.2.1. The online collection 222
   6.2.2. Web site structure 223
   6.2.3. Measurement of web site traffic 223
   6.2.4. Analysis of web site statistics 223
   6.2.5. Limitations 224
6.3. Results **224**
   6.3.1. Total visits to the online collection 224
   6.3.2. Ten most requested PDF files and web pages 225
   6.3.3. Most requested topics 227
6.4. Discussion **228**
   6.4.1. Importance of the *Oltersdorf Collection* 230
   6.4.2. Future capacity of the African food habits web site 231
   6.4.3. Transferability of African foods and food habits 232
6.5. Résumé **236**
6.6. References **237**

DISCUSSION 244

LIST OF TABLES

Page

Table 1	The major online collections on African food habits	21
Table 2	Online data availability by region	45
Table 3	Data extracted from Tanzania, Kenya, Uganda, Zanzibar Island and Pemba Island (1930s-1960s)	47
Table 4	Main dishes of different ethnic groups in Kenya	53
Table 5	Main dishes of different ethnic groups in Tanzania	54
Table 5 (continue)	Main dishes of different ethnic groups in Tanzania	55
Table 6	Main dishes of different ethnic groups in Uganda	56
Table 7	Investigations in North-East Tanzania	99
Table 7 (continue)	Investigations in North-East Tanzania	100
Table 8	Investigations in Kenya	101
Table 8 (continue)	Investigations in Kenya	102-105
Table 9	Investigations in Uganda	106
Table 10	Characteristics of eight dietary assessments and health status surveys on non-communicable disease (NCDs) risk factors	199
Table 10 (continue)	Characteristics of eight dietary assessments and health status surveys on non-communicable disease (NCDs) risk factors	200-211
Table 11	Importance of the *Oltersdorf Collection*	230

LIST OF FIGURES

Page

Figure 1	Primary reason among the ICN interviewees as to why the traditional African diet could be considered healthier	23
Figure 2	Flow of reports of the *Oltersdorf Collection* included/excluded from review	79
Figure 3	*Colonial* and *Neocolonial* contributors to the change in traditional East African food habits	128
Figure 4	Trends of wheat importation in Kenya, Tanzania and Uganda (1962-2004)	138
Figure 5	Main imports of hydrogenated oils into Uganda (1994-2004)	139
Figure 6	Traditional crops available for consumption in Kenya (1961-2003)	141
Figure 7	Traditional crops available for consumption in Uganda (1961-2003)	141
Figure 8	Food crops available for consumption in Tanzania (1961-2003)	142
Figure 9	Flow and characteristics of publications included/excluded for review	180
Figure 10	Total visits to the African food habits web site over 31 weeks	225
Figure 11	Most requested PDF files and/or web pages over 31 weeks	226
Figure 12	Most requested web pages topics among the four country directories over 31 weeks	227
Figure 13	Most requested web pages topics among the four country directories over 31 weeks	228

LIST OF ABBREVIATIONS

AA	Arachidonic Acid
AIDS	Acquired Immune Deficiency Syndrome
AJFAND	African Journal of Food Agriculture Nutrition and Development
AJOL	African Journals Online
APJCN	Asia Pacific Journal of Clinical Nutrition
BMI	Body Mass Index
BP	Blood Pressure
Ca	Calcium
CARDIAC	Cardiovascular Disease and Alimentary Comparison
CCH	Complex Carbohydrates
CHD	Coronary Heart Disease
CIAT	International Center for Tropical Agriculture
CGIAR	Consultative Group on International Agricultural Research
CVD	Cardiovascular Disease
DBP	Diastolic Blood Pressure
DGLA	Dihomo-Gamma-Linolnic-Acid
DHA	Docosahexaenoic Acid
DPA	Docosapentaenoic Acid
EPA	Eicosapentaenoic Acid
FA	Fatty Acids
FAO	Food and Agriculture Organization of the United Nations
FAOSTAT	Food and Agriculture Organization of the United Nations Statistical Database
FDI	Foreign Direct Investment
FFQ	Food Frequency Questionnaire
GATT	General Agreement on Tariffs and Trade
GLV	Green Leafy Vegetables
HB_{A1c}	Glycosylated Haemoglobin
HDL	High Density Lipoprotein
HDLC	High Density Lipoprotein Cholesterol
HIV	Human Immunodefficiency Virus
ICARDA	International Center for Agricultural Research in the Dry Areas
IFIs	International Financial Institutions
IITA	International Institute of Tropical Agriculture
ICN	International Congress on Nutrition
ICRISAT	International Crop Research Institute for the Semi-arid Tropics
IDIFA	Initiative for the Development of Indigenous Food-plants of Africa
IFPRI	International Food Policy Research Institute
IMF	International Monetary Fund
IPGRI	International Plant Genetic Resources Institute
ISNAR	International Service for National Agricultural Research
I.U.	International Units
K	Potassium
LDL	Low Density Lipoprotein
LDLC	Low Density Lipoprotein Cholesterol

Lp(a)	Lipoprotein (a)
MRP	Morogo Research Programme
MUFA	Monounsaturated Fatty Acids
Mg	Magnesium
Na	Natrium
NaCl	Natrium Chloride
NewCROP	New Crops Resource Online Program
NCD	Non-Communicable Disease
NCDs	Non-Communicable Diseases
NR	Not Reported
NR-NCDs	Nutrition Related Non-Communicable Disease
NS	Not Significant
PROTABASE	Plant Resources of Tropical Africa Base
PUFA	Polyunsaturated Fatty Acids
RDI	Recommended Dietary Intake
REE	Resting Energy Expenditure
SAFA	Saturated Fatty Acids
SANBI	South African National Biodiversity Institute
SAFOODS	South African Food Composition Database
SBP	Systolic Blood Pressure
SEPASAL	Survey of Economic Plants for Arid and Semi-Arid Lands
TC	Total Cholesterol
TG	Triglyceride
Total ω-3 FA	Total Omega Three Fatty Acids
Total ω-6 FA	Total Omega Six Fatty Acids
WHO	World Health Organization

The following quote from the *Ngoni* people, a dispersed ethnic group living in East-Central Africa, captures the essence of this dissertation:

The Ngoni themselves say that the change of the diet is the chief cause of their smaller stature today and the prevalence of various illnesses. The diet, by their own account, is much less varied than formerly, and much less meat and milk are consumed.

The knowledge of how to prepare (the special dishes) is passing as the older women die......The old Ngoni dishes, which were of considerable dietetic value, are unknown to the modern mission-trained girls who hanker after wheat flour and sugar which they have seen and used in missionaries' houses, and which can only be bought for money in the stores.

Read M. Native standards of living and African culture change. *Africa.* 1938; 11(Suppl. 3):1-64.

INTRODUCTION AND OBJECTIVES

The purpose of this project was to investigate the health-related importance of traditional African food habits, and advance awareness of the richness of the African food culture. A diversified, traditional diet may be of utmost importance in alleviating current non-communicable disease (NCD) epidemics throughout Africa. Recent NCD epidemics (i.e. diabetes, hypertension, obesity, and cardiovascular disease) appear to be the result of a *nutrition transition* whereby traditional food habits have been progressively replaced by a *globalized food culture*.

The project culminated in the development of an online collection (web site) (which was available at: http://www.healthyeatingclub.com/Africa/ and has been transferred to http://www.drverena.com) which now serves as an important research and educational tool for nutritional scientists and the general public interested in investigating traditional African food habits. The online collection was based upon a unique and precious set of studies obtained through the activities of the *Max-Planck Nutrition Research Unit*, formerly located in Bumbuli, Tanzania. These studies, conducted from the 1930s to 1960s, provide the first empirical evidence of traditional food habits collected from East Africa, including the countries of Tanzania, Kenya, and Uganda. The official caretaker of these studies, Professor Ulrich Oltersdorf, who was also involved in some of the research at the *Max-Planck Nutrition Research Unit*, graciously made the studies available for the purpose of contributing significantly to the development of an online collection. The series of studies have therefore been entitled the *Oltersdorf Collection*

The rationale for the development of the online collection of traditional African foods and food habits was based on: (1) a systematic online search, performed to assess current gaps in online collections, and (2) a questionnaire, administered to opinion leaders in the nutritional sciences at the 18th International Congress on Nutrition (ICN) in Durban, South Africa, in September 2005. This investigation is presented in Chapter 1: *The Need for an Online Collection of Traditional African Food Habits*, published in the *African Journal of Food Agricultural Nutrition and Development (AJFAND)*, 2007 (in press).

A systematic review of the *Oltersdorf Collection* was conducted to extract and collate pertinent data for the online collection. This investigation is presented in Chapter 2: *Content of a Novel Online Collection of Traditional East African Food Habits (1930s - 1960s): Data Collected by the Max-Planck Nutrition Research Unit, Bumbuli, Tanzania*, published in the *Asia Pacific Journal of Clinical Nutrition (APJCN)*, 2007 (in press). This review provided reason to believe that a healthy, diversified traditional diet may have been possible for the indigenous people of East Africa during this period.

To support the contention that traditional African food habits are associated with significant health benefits, a systematic review of the *Oltersdorf Collection* was undertaken to investigate relationships between dietary intake and health status indices within specific cohorts in East Africa (i.e. Tanzania, Kenya and Uganda) from the 1930s to 1960s. This investigation is presented in Chapter 3: *Investigation of the Dietary Intake and Health Status in East Africa in the 1960s: A Systematic Review of the Historic Oltersdorf Collection*, submitted to *Ecology of Food and Nutrition*, November 2006. The review revealed that while non-communicable diseases (NCDs) were not prevalent, there was substantial reporting of malnutrition-related and infectious diseases among many cohorts due to a simplified, monotonous diet during this period.

The findings of the systematic review in Chapter 3 led to the investigation of factors which may have been responsible for the dietary simplification in East Africa during the 1960s. Chapter 4, *Colonial and Neocolonial Forces and the Eradication of Traditional Food Habits in East Africa: Historical Perspective on the Nutrition Transition,* submitted to *Public Health Nutrition*; November 2006, reveals factors which have underpinned the *nutrition transition* in the countries of East Africa (i.e. Kenya, Uganda and Tanzania) from early colonization to the current, oppressive political-economic structure.

Although the *nutrition transition* has affected much of East Africa today, there remain some cohorts which still consume a traditional diet. A systematic review of recent empirical investigations was therefore performed to determine if adherence to a traditional East African diet was associated with better markers of health status, including a lower NCD risk factor profile, versus adherence to a non-traditional diet. This investigation is presented in Chapter 5, *Is the Non-Communicable Disease Epidemic in East Africa being caused by a Loss of Traditional Food Habits?* The studies reviewed provide some support for the health-related benefits of a traditional diet. However, further research is likely required to conclusively demonstrate the association between traditional East African food habits, health status and longevity.

The purpose of Chapter 6, *Utilization and Future Applications for an Online Collection of Traditional African Food Habits,* was to evaluate the overall utilization of the online collection over the first 31 weeks, and to discuss its potential future applications. Overall, visits to the online collection increased by 17%, from week 1 to week 31. On average, our web site accounted for 2958 visits per week. The results suggest that our online collection is increasing in popularity (by 17%), and is frequently accessed for various topics and PDF files. It can be hypothesized that our online collection can successfully contribute to the documentation, compilation, and dissemination of information pertaining to traditional African food habits. The future development of the web site project will include the expansion of data availability of current and historical research data pertaining to traditional foods and food habits throughout all of Africa. In addition, the web site may enhance the development of an online network of communication in and outside of Africa for the creation of targeted and relevant collaborative research projects.

Layout of the book

This book is comprised of the introduction, explanation of definitions, 6 chapters, main discussion and summary. The references (in the style of the Vancouver agreement: *Journal of the American Medical Association*) are quoted at the end of each particular section, to enable a qualitative presentation of the large amount of literature, included in this project.

DEFINITIONS

Food habits

Individuals and cultural groups structure dietary intake according to specific patterns. These patterns include ordinary daily rounds of meals and snacks, as well as annual cycles of feasts and fast days which, in combination with individual foods, comprise *food habits*.[1]

Indigenous food crops and food habits

A food crop or food habit whose natural home is known to be of a specific region (i.e. Tanzania, Zanzibar Island and Pemba Island, Kenya and/or Uganda, in the case of this thesis).[2]

Traditional food crops and food habits

An indigenous or introduced species or food habit which, due to very long use, has become part of the culture of a community.[2]

Community

A group of people living together in a specified area whose members have something in common. In this thesis, the term can be used interchangeably with 'ethnic group'.[2]

References

1. Messer E. *Mehtods for Studying Determinants of Food Intake.* In: Pelto G, Pelto P, Messer E, eds. Research Methods in Nutritional Anthropology. Hong Kong: The United Nations University; 1989:1-201.
2. Maundu P. The Status of Traditional Vegetable Utilization in Kenya. International Plant Genetics Resource Institute (IPGRI). Available at: http://www.ipgri.cgiar.org/publications/HTMLPublications/500/ch09.htm.

1. NEED FOR AN ONLINE COLLECTION OF TRADITIONAL AFRICAN FOOD HABITS

1.1. Background and objectives for the rationale to develop an online collection of traditional African food habits

The Global Burden of Disease study[1] noted that deaths due to non-communicable diseases (NCDs) have increased dramatically in sub-Saharan Africa, and will account for nearly 45% of deaths in the region by 2020. Recent statistics from the World Health Organization revealed nearly 80% of deaths attributable to NCDs worldwide occur in developing countries.[2] This statistic is notable in light of the obesity-diabetes epidemic occurring in developed countries.

Numerous empirical and investigative reports have indicated that current NCD trends in Africa can, in many ways, be attributed to rapid socioeconomic shifts created by an increasi ngly accelerated agenda for global hegemony fueled by western corporate and political interests.[3-5] This agenda, typically identified as *globalization* or westernization, has been implicated in rapid urbanization rates, degradation of the environment, and virtual obliteration of the traditional culture of Africa,[6-8] a continent considered by many to be the *Cradle of Civilization*.

Amongst the difficulties facing the indigenous peoples of Africa today has been the deleterious shift from traditional food habits to processed and packaged food products of western-owned corporations.[9, 10] Consumption of these food products results in elevated intake of saturated fat, trans-fatty acids and food preservatives, and reduced intake of dietary fibre, vital nutrients and phytochemicals when compared to basic dietary guidelines.[11-13] This shift from traditional foods to westernized food products has been dubbed the *nutrition transition*, and has been directly implicated in the rise of type 2 diabetes, CVD, hypertension, obesity, cancer, and related NCDs throughout Africa.[14-16] Moreover, non-communicable, chronic diseases have not simply replaced infectious and malnutrition-related diseases in Africa. Rather, these vulnerable populations now experience a polarized and protracted *double burden* of disease, where the effects of the *nutrition transition* are added to the existing infectious disease burden.[15, 16]

Food habits are amongst the oldest and deeply ingrained aspects of culture. For example, historical evidence of African food habits dating back to the Stone Age has been found in Olorgesailie, Kenya, a historical site on the floor of the Great Rift Valley, approximately 70km south of Nairobi. Over 5,000 years ago hunter-gatherers, commonly called the *ndorobo*, occupied much of East Africa. The *ndorobo* were assimilated by migrants and lost much of their cultural identity, including the loss of knowledge of their food habits.[17] Interestingly, Eaton and Konner[18] investigated dietary shifts over several millennia in Africa and concluded that the hunting and gathering subsistence diet of paleolithic times was superior to the present-day diet largely based on processed and manufactured foods. Throughout history external influences have brought about changes in African food habits, and this has perhaps never been more apparent than the present day.

Food habits are based on traditions, but these traditions change with external influence.[19] The faster people adopt new food patterns, the less likely traditional food knowledge will be passed on to the next generation. In general, the loss of traditional food habits results in a decrease in culture-specific food activities, a decrease in dietary diversity and, if history and current trends are of any indication, significant reductions in economic circumstances, health status, quality of life, and cultural integrity.[20]

Clearly, there is a vital need to investigate and document knowledge of traditional African food habits. This knowledge is necessary to gain an understanding of how traditional dietary patterns could potentially reverse current NCD trends and improve the health status of indigenous populations throughout Africa, and perhaps abroad. Intensive exploration of traditional African food habits could provide insight into the vast and nutrient-rich diversity of foods available in various regions of this vast continent.[12, 21]

Historical, empirical evidence of the richness of traditional African food habits is currently coming to light. Our research group, through Professor Ulrich Oltersdorf,[22] recently gained access to a unique collection of data obtained through the activities of the *Max-Planck Nutrition Research Unit*, previously located in Bumbuli, Tanzania (formerly Tanganyika). This valuable collection provides evidence of the traditional foods and food habits of various ethnic groups located throughout Kenya, Tanzania and Uganda from the 1930s to the 1960s. The collection includes data pertaining to: traditional foods, food taboos, food preparation practices, and agricultural practices, local markets, cooking methods, nutritional status in relation to dietary intake, and chemical composition of traditional foods and their health implications.

This new evidence has the potential to trigger more thorough investigation of traditional African food habits today, and may precipitate the revelation of additional historical knowledge. Moreover, this collection of studies may stimulate the collation of current, original research on traditional African food habits, especially that which is being conducted by indigenous Africans who are currently leading many important investigations. For example, Imbumi *et al.*[23] recently reported on the traditional African food habits of the *Maasai* tribe living in the southern parts of Kenya and the northern district of Tanzania, including traditional staple foods, food preparation practices, food taboos, and changes in dietary patterns over time. At present, the historical dataset collected by the *Max-Planck Nutrition Research Unit* and additional historical and novel sources of information on traditional African food habits, has not been amalgamated and has not been made available for access by researchers and the public.

There is currently a vital need to collect historical and current data on traditional African food habits and present this information via an innovative, online collection. Raising awareness and inspiring investigation of traditional African food habits may be of significant cultural and health-related importance for the indigenous people of Africa as well as the global population at large, given the current NCDs trends sweeping our planet and the potential, positive health-related implications of the traditional African diet.

Therefore, the objectives of the present investigation were: (1) to determine gaps in online collections contributing toward the advancement of knowledge of traditional African food habits, and (2) to determine if opinion leaders in the field of nutritional sciences were aware of the *nutrition transition* and the loss of food culture in Africa, and the potential importance and novelty of creating an online collection of traditional African food habits.

1.2. Methods to investigate the need for an online collection of traditional African food habits

1.2.1 Systematic review of online collections

A systematic search was performed to determine if there was a gap in online collections focused on disseminating information related to traditional African food habits.

Criteria for considering collections

Databases and websites were included in the systematic review to determine whether they contained data or descriptions of traditional African food habits, including: traditional staple foods, food balance sheets, dietary practices (e.g. preparation, cooking techniques, and flavoring), food taboos and customs, chemical composition of traditional African foods, and classification systems of individual foods (e.g. staple foods, green leafy vegetables, roots and tubers).

Search method

The review of online databases and websites was conducted between June 2005 and April 2006, and limited to the English language. The search combined key words: African food habits, Africa, traditional foods, indigenous foods, diets, crops, wild species, food culture, dietary practices, information networks, information systems, databanks, databases, libraries, and involved:

(1) A systematic search of primary internet search engines: Google, Yahoo, and AltaVista;

(2) A systematic search of computerized databases: Web of science and Ovid;

(3) A systematic search of site-specific engines of the following

organizations: Food and Agriculture Organization of the United Nations (FAO), World Health Organization (WHO), International Food Policy Research Institute (IFPRI), International Plant Genetic Resources Institute (IPGRI), World Vegetable Center, Consultative Group on International Agricultural Research (CGIAR), International Center for Tropical Agriculture (CIAT), International Center for Agricultural Research in the Dry Areas (ICARDA), International Crop Research Institute for the Semi-Arid Tropics (ICRISAT), International Institute of Tropical Agriculture (IITA), International Service for National Agricultural Research (ISNAR) and the South African National Biodiversity Institute (SANBI).

In addition, web sites, databases and articles retrieved were examined for further relevant links and references.

1.2.2. Questionnaire administration at the 18th International Congress of Nutrition (ICN)

A questionnaire was administered to opinion leaders in the nutritional sciences at the 18th ICN in Durban, South Africa, September 2005 (Appendix A). The purpose was: (1) To determine if there was awareness of the importance of traditional African food habits within the context of the *nutrition transition* currently plaguing Africa, and (2) To determine if there was general support among this cohort for an online collection of traditional African food habits.

The questionnaire mode included both open and closed format (Yes/No) questions, and was divided into three sections. Section I included standard demographic questions. Section II included questions on the *nutrition transition*, the loss of traditional food habits and related outcomes in Africa, which included a special subsection completed by indigenous African opinion leaders regarding past and present staple foods from their region of origin. Section III included questions regarding the importance of amalgamating and providing data on traditional African food habits via an online collection. Participants were shown a sample of potential web pages displayed offline, using a laptop computer and the appropriate software (Explorer ProTM MetaProducts).

The principal investigator distributed and collected all questionnaires. Primary responses from the three sections were analyzed using SPSS software for Windows, Release 11.0.0 (SPSS Inc, Chicago, Illinois).

1.3. Results

1.3.1. Systematic review of online collections on African food habits

The systematic review of online collections on traditional African food habits resulted in nine collections being identified. These included: one food supply database and two food composition databases,[24-26] two databases providing information on "wild" and semi-domesticated plants of tropical and subtropical drylands, including Africa,[27, 28] two databases providing information on various crop species,[29, 30] and two online-publication catalogue databases.[31, 32] The nine online databases with a focus on African food habits are described in Table 1.

Table 1: The major online collections on African food habits.

Database (time series)	Available data related to African food habits
AJOL[31] 1998 - present	Database of 230 African-published journals
IPGRI-online publications catalogue/database[32] 1977 – present	Publications on indigenous/traditional food systems and documentation about indigenous species/foods for all of Africa
FAOSTAT[24] 1961 – present	Food balance sheets for all African countries, including information on: -Domestic supply -Domestic utilization -Per capita supply
AFROFOODS*[25] 1952 – 1999	Food composition table for various African countries
SAFOODS[26] Published in 1991	Energy, macronutrient composition of different foods consumed by people in South Africa
SEPASAL[27] since 1981	Information on more than 6300 useful wild and semi-domesticated plants of tropical and subtropical drylands including Africa (scientific name, plant family, geographical distribution, ecology, use of plants, properties and chemical analysis)
PROTABASE[28] initiated in 2000	Review articles for nearly 400 African plant species (botanical names and botanical descriptions of their useful properties, cultivation and potential as a crop)
NewCROP[30] since 1995	Crop database on scientific crop profiles including African species (crops by name, uses including food use, geography, commodity, cultural practices, nutritional value)
Famine food database[29] since 1995	Information on African crops (scientific name, plant family, vernacular, geographical distribution, ecology, use and preparation of plants)

*one of INFOODS (International Network of Food Data Systems) regional data centers

1.3.2. Questionnaire responses from opinion leaders in the nutritional sciences

Participants

Ninety-two questionnaires were completed at the ICN in Durban, South Africa 2005. Mean age of the interviewees was 36.3 ± 9.7 years. All participants had completed tertiary education in the nutritional sciences, with minimum attainment of a Masters degree. The majority (66%) reported Africa as their continent of residence, including 62% from Southern Africa (i.e. South Africa, Angola, Botswana, and Malawi), 15% from West Africa (i.e. Mali, Ghana, and Nigeria), 13% from East Africa (i.e. Uganda, Ethiopia, and Kenya), 5% from Central Africa (i.e. Cameroon, Republic of Congo), and 5% from North Africa (i.e. Egypt).

Awareness of the nutrition transition

Approximately 86% of respondents agreed that a *nutrition transition* from a traditional to a westernized diet is currently affecting urban sub-Saharan Africa. Only 14% of the interviewees disagreed on the occurrence of the *nutrition transition*.

Past and present staple foods

Approximately 62% of indigenous African interviewees identified maize as a primary staple food of the past within their region of origin. Rice (25%), legumes (25%), green leafy vegetables (22%), roots and tubers (22%), meat and poultry (20%), sorghum (17%), millet (17%), and plantains (14%) were also mentioned as past staple foods.

When questioned about present staple foods of their region of origin, there were notable reductions in the identification of: sorghum (declined to 0%); millet (declined to 8%), green leafy vegetables (declined to 9%), and legumes (declined to 14%).

Comparison of past and present diet

The majority of interviewees (84%) believed that the traditional African diet was healthier than the current westernized diet. Primary reasons provided as to why the traditional diet could be considered healthier are presented in Figure 1.

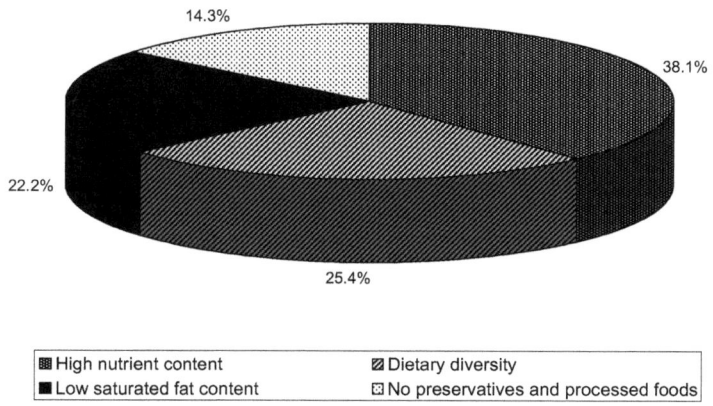

Figure 1. Primary reasons among the ICN interviewees as to why the traditional African diet could be considered healthier.

Tb traditional African foods most commonly associated with health benefits, as identified by the interviewees, included: millet, green leafy vegetables, roots and tubers, fruits, legumes, palm oil, wild "bush" meat, and maize.

On adherence to the traditional African diet, 52% of the participants agreed that the majority of people in Africa (rural and urban) still eat the traditional African diet.

Factors responsible for the nutrition transition and double burden

Primary factors responsible for nutritional deficits in Africans today were identified by the interviewees as: low nutritional value of the current westernized diet (39%); economic pressures related to westernization/*globalization* (32%); and reduced availability and access to quality foods (such as scarcity through lack of options) (25%).

The major factors contributing to the *double burden* epidemic as noted by the cohort, included: urbanization, associated economic pressure and maldistribution of wealth (33%), adoption of western cultural beliefs (17%), adoption of an unhealthy monotonous diet including excessive energy consumption in urban areas and under-nutrition in rural areas (16%). Several interviewees identified lack of available infrastructure (14%), including lack of basic healthcare, loss of arable land/habitat, loss of biodiversity and reduced access to quality foods as the major causes of the *double burden*. Existing disease burden (7%) and lack of knowledge of what is considered as 'healthy diet' (7%) were also identified as influential factors.

Loss of knowledge of traditional African food habits

The majority of the interviewees (78%) believed that knowledge of traditional African food habits is being lost. Approximately 56% believed that the lack of promotion, documentation and research of indigenous foods in Africa was the main source for the loss of knowledge. Other reasons identified included westernization/*globalization* and/or colonization (35%), ignorance and stigma of traditional food (5%) and international food aid programs (5%).

Importance of an online collection of traditional African food habits

Open-ended questioning revealed that the online collection could serve as an important research and educational tool (70%). Several interviewees commented that an online collection could support preservation of knowledge of African food habits (17%) including their potential health implications (13%).

The need for an online collection

The opinion leaders were virtually unanimous (88%) in suggesting that an online collection of traditional African food habits should be used for educational purposes.

The vast majority of opinion leaders (82%) believed that a gap currently exists in online empirical evidence related to traditional African food habits. The majority of respondents (69%) were not aware of scientists currently investigating traditional African food habits. The vast majority of interviewees (92%) indicated that they would make use of a novel online collection, if made available.

1.4. Discussion

The investigation revealed several important findings that support our proposal for an innovative, online collection of traditional African food habits. The systematic review revealed nine online databases that provide some data pertaining to certain aspects of traditional African foods (Table 1). All of these collections have important implications, but fundamentally differ from our current vision of an online collection of traditional African food habits designed to stimulate education and research of food habits and their health implications, and provide a well-rounded forum in which such information can be presented and shared. According to the systematic search, there are currently no online collections that have an overall focus on traditional African food habits. Moreover, 82% of the opinion leaders at the 18th INC 2005 in Durban, South Africa believed that a gap currently exists in this area.

Overwhelmingly, the opinion leaders surveyed believed that the traditional African diet was superior to the increasingly prevalent westernized diet, citing 'nutrient density and diversity', 'low saturated fat' and 'no preservatives' as key determinants of health. Moreover, the indigenous African experts interviewed noted reduced millet and sorghum consumption and increased wheat and rice consumption as primary staple foods within their regions of origin. Empirical investigations have demonstrated the superior nutritional indices of millet and sorghum as compared to rice and wheat.[33-38] The notable finding provides one example of how staple foods in Africa are shifting toward an unfavorable direction.

According to the interviewees, the adoption of western values, urbanization, economic pressures, maldistribution of wealth, and scarcity through lack of choice were primary factors driving the *nutrition transition* and the related *double burden* epidemic in Africa today. The increasing prevalence of NCDs associated with these socio-economic pressures has been well described.[9, 12, 39-41] Loss of cultural ties, traditional knowledge and traditional food resources occur with urbanization.[42] Moreover, Bourne et al.[4, 40] reported that in South Africa the westernization of diet is occurring in rural areas, and is not only confined to urban centers.

It is essential to provide data on the nutrient and dietary intake of Africans prior to the onset of the *nutrition transition*.[6] According to Popkin,[43] the increased consumption in refined foods and fats among urban

Africans is due to the appearance of dietary shifts. The occurrence of a dietary shift was also highlighted in the stakeholder survey. Documentation and presentation of these new dietary patterns of Africans couldbe integrated into the vision of an online collection, potentially providing further support for the traditional African diet. Comparative data on food quality and health status during periods of transition may enhance advocacy for the traditional diet amongst health care, and nutrition intervention programs throughout Africa, and perhaps abroad, in countries drawing African migrants and refugees.

Globalization of culture and commercial activities promulgates a westernization of developing-country food systems and diets.[44] The complex set of industrial and modernizing influences involved lead to a delocalization of food supply, which has been described as a major determinant of dietary change.[45] With the increasing networks of socioeconomic and political interdependencies, a decreasing diversity of food items are consumed.[20] The investigative group, including the principle investigator (V. R.) and four co-investigators in the field of nutritional sciences (U. Oltersdorf, I. Elmadfa, M.L. Wahlqvist and A. Kouris-Bazos), believes that indigenous knowledge regarding food choices should be amalgamated with historical empirical knowledge and novel scientific investigation of the chemical composition of foods, including nutrients and non-nutrients (fibers, polyphenols, etc.). This combination of indigenous and scientific knowledge may increase the marketability of traditional African food items. For example, previous marketing of culture-specific food items via traditional knowledge and scientific inquiry have led to increased advocacy and popularity of Mediterranean and Japanese cuisines.

The high quality and diversity of the traditional African diet was noted by early European travellers.[46, 47] For example, Livingstone was surprised to see such a variety of foods eaten by the *Wagogo* people in Central Tanzania.[46, 47] The majority of the interviewees shared the opinion that the traditional African diet was healthier due to its high nutrient content, high diversity, low saturated fat content and the absence of preservatives. The sensory and culinary properties of local food crop varieties, the diversity of the foods used and the potential genetic variations in nutrient composition within neglected and under-utilized species are further examples of the type of information which should be presented via an online collection.

The diversity of indigenous crops, wild plants and animal species available in most tropical countries, in addition to providing essential nutrients, presumably offer health benefits.[48, 49] Several empirical investigations have associated traditional African items with health benefits, including various species of green leafy vegetables,[50] grain legumes,[51] palm fruit[52, 53] and millets.[34] These food items were identified as healthy by some of the interviewees at the 18th ICN.

In summary, the investigation revealed a clear need for a novel, online collection of traditional African food habits. This collection could serve as an important medium for education, research, and international networking, according to experts in the nutritional sciences surveyed at the recent ICN. The majority of the respondents believed that knowledge of traditional African food habits is being lost, and that they would make use of an online collection on traditional African food habits, if available.

1.5. Résumé

Amongst the difficulties facing the indigenous people of Africa today is the deleterious shift from traditional food habits to the processed and packaged food products of western-owned corporations. This *nutrition transition* has been implicated in the rise of non-communicable diseases (NCDs) throughout Africa. The purpose of the present investigation was to determine whether there is a current need to document traditional African food habits via an online collection in an attempt to stimulate further research in this area and potentially improve the health status of indigenous Africans threatened by the *nutrition transition*. A systematic search was performed to assess possible gaps in online collections focused on traditional African food habits. A questionnaire was administered to opinion leaders in the nutritional sciences at the 18th International Congress of Nutrition (ICN) in Durban, South Africa, September 2005, to determine the level of awareness of the importance of traditional African food habits within the context of the *nutrition transition*, and to determine the support among this cohort for an online collection of traditional African food habits. The systematic review resulted in nine collections being identified. None of these collections however, were specifically designed to raise awareness of traditional African food habits. Findings from the survey revealed that 86% of the cohort agreed that Africa is currently undergoing a *nutrition transition*. Nearly 80% believed that knowledge of traditional African food habits is being lost. Indigenous African interviewees noted reduced consumption of sorghum and millet and an increased consumption of wheat and rice within their region of origin. Approximately 82% believed that there was currently a gap in online collections focused on presenting information on traditional African food habits. Ninety-two percent of the cohort indicated their preparedness to make use of an innovative, online collection of data on traditional African food habits. The findings revealed a critical need to collate and present data on traditional African food habits via an online collection that could be used to stimulate education and research of food habits and their health implications, to provide a well-rounded forum in which such information is presented and shared.

1.6. References

1. Murray C, Lopez A. Mortality by cause for eight regions of the world: Global burden of disease study. *Lancet.* 1997; 349:1269-1276.
2. World Health Organization. Diet, physical activity and health. Fifty-five World Health Assembly. Geneva: World Health Organization; 27 March 2002:1-5.
3. Bovet P, Ross A, Gervasoni J, et al. Distribution of blood pressure, body mass index and smoking habits in the urban population of Dar es Salaam, Tanzania, and associations with socioeconomic status. *Int J Epidemiol.* Feb 2002; 31(1):240-247.
4. Bourne LT, Lambert EV, Steyn K. Where does the black population of South Africa stand on the nutrition transition? *Public Health Nutr.* 2002; 5(1A):157-162.
5. Sande M. Cardiovascular disease in sub-Saharan Africa: A disaster waiting to happern. *Neth J Med.* 2003; 61(2):32-36.
6. Popkin BM. Part II. What is unique about the experience in low- and middle-income less-industrialised countries compared with very-high-income industrialised countries? The shift in stages of the nutrition transition in the developing world differs from past experiences! *Public Health Nutr.* 2002 ;5(1A):205-214.
7. Voster H, Bourne L, Venter C, Oosthuizen W. Contribution of nutrition to the health transition in developing countries: A framework for research and intervention. *Nutr Rev.* 1999;

57(11):341-349.
8. Guest R. *The Shackled Continent: Africa's Past, Present and Future.* London: Macmillan; 2004.
9. Maletnlema T. A Tanzanian perspective on the nutrition transition and its implication for health. *Public Health Nutr.* 2002; 5(1A):163-168.
10. Popkin BM, Lu B, Zhai F. Understanding the nutrition transition: measuring rapid dietary changes in transitional countries. *Public Health Nutr.* 2002; 5(6A):947-993.
11. Shetty P. *Diet Nutrition and Chronic Disease: Lessons from Contrasting Worlds.* Vol 15-16. Chichester, UK: John, Wiley & Sons; 1997.
12. Drewnowski A, Popkin B. The nutrition transition: New trends in the global diet. *Nutr Rev.* 1997; 55(2):31-43.
13. Walker A, Segal I. Health/ill-health transition in less privileged populations: What does the future hold? *J R Coll Physicians Lond.* 1997; 31(4):392-395.
14. World Health Organization (WHO). Globalization, diet and noncommunicable disease. Geneva 2002:1-185.
15. Popkin B. An overview of the nutrition transition and its health implications: The Bellagio meeting. *Public Health Nutr.* 2002; 5:93-103.
16. Popkin BM. The nutrition transition and prevention of diet related disease in Asia and the Pacific. *Food Nutr Bull.* 2001; 22:1-58.
17. Maundu P, Imbumi M. East Africa. In: Katz S, Weaver W, eds. *Encyclopedia of Food and Culture.* Vol 3. New York: Thomson and Gale; 2003:27-34.
18. Eaton S, Konner M. Paleolithic nutrition. *N Engl J Med.* 1985;

312(5):283-290.

19. Oniang'o RK. Food habits in Kenya: The effects of change and attendant methodological problems. *Appetite.* 1999; 32:93-96.

20. Kuhnlein H, Receveur O. Dietary change and traditional food systems of indigenous peoples. *Annu Rev Nutr.* 1996; 16:417-442.

21. Popkin B. The nutrition transition and its health implications in lower income countries. *Public Health Nutr.* 1998; 1:5-21.

22. Oltersdorf U. Comparison of nutrient intakes in East Africa. Paper presented at: *The Human Biology of Environmental Change.* 5-12 April, 1971; Blantyre, Malawi. 1971: 51-59.

23. Imbumi M, Saitabu H, Maundu P. Maasai traditional foods: A look at diets in the Maasai culture. *Ann Nutr Metab.* September 2005; 49(Suppl. 1):381.

24. Food and Agriculture Organization of the United Nations (FAO). FAOSTAT, FAO Statistical Databases On-line. Available at: http://faostat.fao.org/. (Accessed 24 March, 2006).

25. Food and Agriculture Organization of the United Nations FAO. International Network of Food Data Systems (INFOODS) On-line. Agriculture Biosecurity Nutrition and Consumer Department. Available at: http://www.fao.org/infoods/tables_africa_en.stm. (Accessed 08 April, 2006).

26. Medical Research Council (MRC) South Africa. (SAFOODS) South African Food Composition Database On-line. Medical Research Council (MRC) South Africa. Available at: http://www.mrc.ac.za/FoodComp/. (Accessed 26 March, 2006).

27. Royal Botanic Gardens Kew. Survey of Economic Plants for Arid and Semi-Arid Lands (SEPASAL) On-line. Royal Botanic Gardens Kew, Centre for Economic Botany (CEB). Available at:

http://www.rbgkew.org.uk/ceb/sepasal/internet/. (Accessed 27 March, 2006).

28. Prota Foundation. PROTABASE (Plant Resources of Tropical Africa) On-line. Prota Foundation 2006. Available at: http://database.prota.org/search.htm. (Accessed 07 April, 2006).

29. Freedman R. The Famine Food Database Online. Purdue University Center for New Crops and Plant Products. Available at: http://www.hort.purdue.edu/newcrop/faminefoods/ff_home.html. (Accessed 07 April, 2006).

30. Purdue University Center for New Crops and Plant Products. NewCROP (New Crops Resource Online Program). Center for New Crops and Plant Products, Purdue University, Department of Horticulture and Landscape Architecture. Available at: http://www.hort.purdue.edu/newcrop/default.html. (Accessed 07 April, 2006).

31. African Journals Online (AJOL). African Journals Online. NISC SA (National Inquiry Services Centre). Available at: http://www.ajol.info/. (Accessed 26 March, 2006).

32. International Plant Genetic Resources Institute (IPGRI) On-line. Pubications-Catalogue. International Plant Genetic Resources Institute. Available at: http://www.ipgri.cgiar.org/system/page.asp?frame=publications/indexpub.htm. (Accessed 24 March, 2006).

33. Dicko M, Hilhorst R, Gruppen H, et al. Comparison of content in phenolic compounds, polyphenol oxidase, and peroxidase in grains of fifty sorghum varieties from Burkina Faso. *J Agric Food Chem.* 2002; 50(13):3780-3788.

34. Nishizawa N, Shimanuki S, Fujihashi H, Watanabe H, Fudamoto Y,

Nagasawa T. Proso millet protein elevates high plasma level of high-density lipoprotein: A new food function of proso millet. *Biomed Environ Sci.* 1996; 9:209-212.

35. Gooneratne J, Munasinghe L, Senevirathne W. Millet bran and corn bran lowers plasma total and LDL cholesterol levels in hypercholesterimic subjects. *Ann Nutr Metab.* 2005; 49(Suppl. 1):304.

36. Kurup P, Krishnamurthy S. Glycemic response and lipemic index of rice, raggi and tapioca as compared to wheat diet in human. *Indian J Exp Biol.* 1993; 31:291-293.

37. Hulse J, Laing E, Pearson O. *Sorghum and Millets: Their Composition and Nutritive Value.* New York: Academic Press; 1980.

38. Food and Agriculture Organization of the United Nations (FAO). Sorghum and millets in human nutrition. Vol 27. Rome: Food and Agriculture Organization of the United Nations (FAO); 1995. Available at: http://www.fao.org/DOCREP/T0818e/T0818e00.htm

39. Levitt N, Katzenellenbogen J, Bradshaw D, Hoffman M, Bonnici F. The prevalence and identification of risk factors for NIDDM in urban Africans in Cape Town, South Africa. American Diabetes Association. *Diabetes Care.* 1993; 16(4):601-607.

40. Bourne LT, Langenhoven ML, Steyn K, et al. Nutritional intake of the African population of the Cape Peninsula, South Africa: The BRISK study. *Cent Afr J Med.* 1993; 39:238-247.

41. Trowell H. From normotension to hypertension in Kenyans and Ugandans 1928-1978. *East Afr Med J.* 1980; 57(3):167-173.

42. Kuhnlein HV, Johns T. Northwest African and Middle Eastern food and dietary change of indigenous peoples. *Asia Pac J Clin Nutr.* 2003; 12(3):344-349.

43. Popkin B. The nutrition transition in low-income countries: An emerging crisis. *Nutr Rev.* September 1994; 52(9):285-298.
44. Cannon G. Nutrition: The new world disorder. *Asia Pac J Clin Nutr.* 2002; 11 (Suppl):498-509.
45. Pelto G, Pelto P. Diet and delocalization: Dietary changes since 1750. *J. Interdis Hist.* 1983; 14(2):507-528.
46. Schaffer R, Finklestein F. *The Food and Growth of Gogo Children.* Stenciled Paper. 1963:1-10.
47. Jelliffe D, Jelliffe E, Benett F, White R. Ecology of childhood disease in the Karamojong in Uganda. *Arch Environ Health.* 1964; 9:25-36.
48. Haltloy A, Hallund J, Diarra M, Oshaug A. Food variety, socioeconomic status and nutritional status in urban and rural areas in Koutialia (Mali). *Public Health Nutr.* 2000; 57:57-65.
49. Ogle B, Hung P, Tuet H. Significance of wild vegetables in micronutrient intakes of women in Vietnam: An analysis of food variety. *Asia Pac J Clin Nutr.* 2001; 10:21-30.
50. James A, Chweya J, Eyzaguirre P. *The Biodiversity of Traditional Leafy Vegetables.* Rome: International Plant Genetic Resources Institute (IPGRI); 1999:1-182.
51. Zulet M, Macarulla M, Portillo M, Noel-Suberville C, Higueret P, Martinez J. Lipid and glucose utilization in hypercholesterolemic rats fed a diet containing heated chickpea (*Cicer aretinum L.*). *Int J Vitam Nutr Res.* 1999; 69:403-411.
52. Wattanapenpaiboon N, Wahlqvist M. Phytonutrient deficiency: the place of palm fruit. *Asia Pac J Clin Nutr.* 2003; 12(3):363-368.
53. Solomons N, Orozco M. Alleviation of vitamin A deficiency with palm fruit and its products. *Asia Pac J Clin Nutr.* 2003; 12(3):373-

384.

1.7. Appendix A

Questionnaire on African Food Habits

This questionnaire will collate your opinions about the development of a web site on FOOD HABITS IN AFRICA and its future importance.

SectionI:
1. Gender:
   - Male
   - Female
2. Country of residence:

3. Age:

4. Highest level of education:

Section II:
Subsection for opinion leaders indigenous to Africa
Which region/province are you from:
1. The main staple food items in my home region were... (list below)

2. Today, the main staple food items in my home region are... (list below)

General questionnaire: for all 18th ICN opinion leaders

3. Do you think that the traditional African diet was healthy?
   - If yes, why:
   - No

4. Do you think the *new, westernized* African diet is better compared to the traditional African diet?
   - Yes
   - No

5. Do you think that the majority of people in Africa still eat the traditional African diet?
   - Yes
   - Not

6. I think the major problems of the African diet today are………………..
   (3 entries are possible)
   a.)
   b.)
   c.)

7. Is the *nutrition transition** happening throughout Africa?
   - Yes
   - No

*adverse dietary shift (e.g. shifts in structure of the diets towards a greater role for higher fat, added sugar foods, reduced fruit and vegetable intake, reduced fibre intake, greater energy density and greater saturated fat intake) which is dominated by nutrition related non-communicable disease (NR-NCDs)

8. The major causes of the *double burden*** in Africa today are…………….
   (3 entries are possible)
   a.)
   b.)
   c.)

**Co-existense of under nutrition, infectious disease with nutrition related non-communicable disease (NR-NCDs) such as for example high blood pressure, obesity, type II diabetes mellitus

9. What are the main reasons for a change in traditional African food habits?
   (3 entries are possible)
   a.)
   b.)
   c.)

10. Do you think that the traditional knowledge of African food habits is being lost?
    o Yes,
       because..
       ....
    o No

Section III: Web site project on East African food habits

11. What do you think about my project, of making baseline information on East African food habits from the 1930s to 1960s available via the internet?

12. The main focus on the web site should be:

13. The knowledge of African food habits is important because:

14. Does a gap on online empirical and precise data on African food habits exist?
    o Yes
    o No

15. Would you make use of the web site?
    - ○ Yes
    - ○ No

16. Are you aware of scientists who study African food habits?
    - ○ If yes, can you name them and provide a contact address?

    ....

    - ○ No

17. Are you interested in more information about the project?
    - ○ Yes
    - ○ No

18. Are you interested in receiving an electronic newsletter about the project news?
    - ○ Yes,

    My e-mail address:............................@..........................

    - ○ No

Copyright©2005
Mag. Verena Raschke
18th International Congress on Nutrition 2005
Durban, South Africa

2. CONTENT OF AN INNOVATIVE ONLINE COLLECTION OF TRADITIONAL EAST AFRICAN FOOD HABITS (1930s-1960s): DATA COLLECTED BY THE *MAX-PLANCK NUTRITION RESEACH UNIT*, BUMBULI, TANZANIA

2.1. Background and objectives for the development of an online collection of traditional East African food habits

Over the past several decades, sub-Saharan Africa has been experiencing a *nutrition transition* whereby traditional foods and food habits have been progressively replaced by the *globalized* food culture of the multinational corporations.[1] The impact has been disastrous. The *nutrition transition* has been directly implicated in the recent upsurge of non-communicable diseases (NCDs) throughout sub-Saharan Africa. The World Health Organization has recently revealed that NCDs currently account for 40% of deaths in developing countries, and this proportion is expected to increase significantly in the years ahead.[2] Within the next twenty years, sub-Saharan Africa can expect a three-fold increase in deaths due to cardiovascular disease (CVD),[3] and a near three-fold increase in the incidence of type 2 diabetes.[4]

Investigations conducted in Okinawa Japan,[5,6] the Mediterranean,[7,8] and China,[9,10] have provided robust evidence that traditional foods and traditional food habits are inextricably linked to vitality and longevity. To gain an insight into the factors potentially responsible for increased quality and quantity of life among these cohorts, it is therefore essential to evaluate the commonalities of their respective cuisines.

Fundamentally, the foods and food habits of these cultural groups overlap with regard to: (1) the utilization of fresh, whole foods, prepared according to traditional, often ancient, practices (2) the absence of corporate influence, which includes the lack of genetically engineered foods, highly processed foods, trans-fatty acids, preservatives, and common excitotoxic additives (e.g. aspartame, monosodium glutamate), some of which are known to induce metabolic abnormalities and hasten the genesis of obesity-related disorders.[11]

Historically, food habits flourish as an understanding of the food environment and the relationship between food choice and health status improves. Cultural beliefs and cultural practices also influence food habits; however health and survival inevitably remain at the forefront of food choice.

For example, Johns et al.[12] recently reported that the *Maasai*, who live in the northern district of Tanzania and the southern parts of Kenya, routinely eat almost double the recommended dietary intake of animal fats, yet their CVD risk remains negligible.[12,13] This paradox may be partially explained by the fact that the *Maasai* are extremely active, and consume a diversified diet including over twenty-five local plant species that contain antioxidants more powerful than vitamins C and E.[12]

The complete eradication of the corporate (i.e. political) domination of traditional foods and food-growing resources may be of utmost importance in averting projected NCDs trends and alleviating malnutrition in Africa. With respect to NCDs, the trends in Africa, and indeed the whole world in general, have been driven by the *scarcity-through-abundance* philosophy of the multinational corporations, which can be summarized as: a lack of quality food choices (scarcity) amongst the massive, insidious web of available options (abundance).

Unfortunately, today, the corporate masters of the so-called *New World Order* and their agenda for global hegemony, have largely succeeded in creating a *globalized* food culture that has been invariably linked to dire health consequences, including diabetes, obesity, CVD, and various cancers. This *globalized* food culture undeniably stands in marked contrast to the food culture Hippocrates spoke of when he stated: "Let food be your medicine and let medicine be your food." Global statistics on the incidence and prevalence of NCDs speak for themselves.

African culture has been, and continues to be, systematically extirpated. This extirpation includes the loss of traditional food habits. The faster people adapt to the *New World, globalized* food patterns, the less likely traditional knowledge will be transferred to the next generation.[14] Inevitably, the loss of knowledge leads to reduced culture-specific food activity, reduced dietary diversity, malnutrition and/or NCDs, and reduced cultural morale.[15]

According to the survey conducted at the 18th International Congress of Nutrition (ICN) in Durban, South Africa, 2005, experts in the nutritional sciences (n=92) were unanimous (84%) in believing that traditional African foods and food habits were superior to the globalized food habits currently underpinning the *nutrition transition*. Further, these experts believed that knowledge of traditional African food habits is being lost, and that there is a critical need for documentation.

With a strong rationale for initiating a project aimed at preserving knowledge of traditional foods and food habits in Africa, we conceived the idea of collating data for an online collection (available at: www.healtheatingclub.com/Africa/).[16] The online collection currently presents information pertaining to traditional foods and food habits of East Africa (i.e. Tanzania, including Zanzibar Island and Pemba Islands, Kenya, and Uganda). These data were amalgamated by reviewing a series of observational studies collected by the *Max-Planck Nutrition Research Unit*, formerly located in Bumbuli, Tanzania. This unique and precious collection of studies has been stored at the Federal Research Centre for Nutrition and Food (BfEL) in Karlsruhe, Germany, and has remained largely inaccessible to researchers and the public. The official caretaker of these studies, Professor Ulrich Oltersdorf, who was also involved in some of the research at the *Max-Planck Nutrition Research Unit*, graciously made the collection available to our investigative team with the purpose of contributing significantly to the innovative project.[17] This series of studies has therefore been entitled the *Oltersdorf Collection*.

The purpose of this chapter is to present the review of the *Oltersdorf Collection* with the primary intent of extracting data for the online collection of traditional African foods and food habits. The specific objectives in reviewing this historical collection were four-fold:

(1) To systematically categorize and extract data pertaining to traditional African foods and food habits

(2) To provide a general overview of these data, with specific emphasis on traditional staple foods, and food preparation practices

(3) To discuss the health implications of these traditional foods and food habits

(4) To propose areas for further investigation and documentation.

2.2. Methods used for the development of the online collection

2.2.1. The Oltersdorf Collection

The *Oltersdorf Collection* consists of 75 observational reports of nutritional outcomes collected by the *Max-Planck Nutrition Research Unit*, in Bumbuli, Tanzania. The investigations were conducted throughout Kenya, Uganda, and Tanzania, including Zanzibar Island and Pemba Island from the 1930s to 1960s. The entire collection of documents has been scanned and converted into PDF files, which are now available for free download.[16]

2.2.2. Data extraction and classification

The 75 reports of the *Oltersdorf Collection* were investigated by the principal researcher (V.R.). All co-investigators were consulted in creating an appropriate classification system. The co-investigators include 4 experts in the field of nutritional sciences (U. Oltersdorf, I. Elmadfa, M.L. Wahlqvist, and A. Kouris-Blazos). Relevant data were classified as follows:

1. Food availability data
2. Chemical composition of foods
3. Staple foods, including native crops, cereals, legumes, roots and tubers, vegetables, fruits, spices, oils/fats, traditional drinks, and animal foods
4. Food preparation and culture, including traditional dishes, food taboos and rituals, cooking methods and preparation, food habits among women, infants and children, agricultural practices, and local markets
5. Dietary intake and health status indicators

2.2.3. The online collection

The web site on African food habits can be entered via the index page which was available at: http://www.healthyeatingclub.com/Africa/ and has been transferred to http://www.drverena.com. The overview contents web page provides a general overview of the online collection, including the introduction, background and aims of the web site PhD project. In addition, the overview contents web page also provides data related to food availability and the chemical composition of foods in East Africa during this period.

Region-specific web pages provide the data related to foods, food habits, and dietary intake and health status indicators within various ethnic groups in Tanzania, Kenya, Uganda, and the Zanzibar Island and Pemba Island Availability of online data by region is presented in Table 2.

Table 2. Online data availability by region.

Web page topic	Country			
	Tanzania	Kenya	Uganda	Zanzibar Island and Pemba Island
Literature content by ethnic groups	X	X	X	
Food balance sheets (1950-1962)	X	X	X	
Chemical compositions of traditional African foods	X	X	X	X
Nutrition transition	X	X	X	
Foods and beverages				
Staple crops	X	X	X	
Cereals	X	X	X	
Legumes	X	X	X	
Root and tubers	X	X	X	
Vegetables	X	X	X	X
Fruits	X	X	X	X
Spices	X	X	X	X
Oils and fats	X		X	X
Traditional drinks	X	X	X	
Animal foods	X	X	X	X
Food habits				
Diet and dishes	X	X	X	X
Taboos and ritual foods	X	X	X	
Cooking methods and preparation	X	X	X	X
Women	X	X	X	
Children	X	X	X	
Agriculture	X	X	X	
Local markets	X	X	X	
Nutrients and health				
Calories	X	X	X	X
Protein	X	X	X	X
Vitamins and minerals	X	X	X	
Health and disease	X	X	X	

2.3. Content of the online collection

Food availability

Food availability data (i.e. food balance sheets) have been collected, and are available for Tanzania,[18, 19] Kenya[18] and Uganda.[20-22]

Chemical composition of foods

Chemical composition data has been presented in eight publications.[23-30]

Staple foods

Investigations documenting traditional food habits in Tanzania, Kenya, Uganda, and the Zanzibar Island and Pemba Island are summarized in Table 3.

Crops

During the 1960s, plantains were a common staple crop around the Lake Victoria region of Uganda, and in the west and Kilimanjaro regions of Tanzania.[31-33] Millet was common in the eastern and northern parts of Uganda, in the Nyanza region of Kenya, and the south side of Lake Victoria up to the Central Region in Tanzania.[34-37] The remaining regions produced maize as a primary staple crop, including the West Nile region in Uganda, the Rift Valley, the Central Region, large sections of the Eastern Region of Kenya, and a belt which stretches from the Pare and Usambara mountains in the north to the central parts of Tanzania. Rice was grown along the coast, on the islands, and in some riverine areas such as Tana in Kenya and Rufiji in Tanzania. Cassava also played an important role in many parts, though primarily as a reserve food.[33, 38-40]

Table 3. Data extracted from Tanzania, Kenya, Uganda, Zanzibar Island and Pemba Island (1930s-1960s).

I. Traditional foods and beverages	
• **Staple crops** Tanzania[31-38] Kenya[39-42] Uganda[43-46]	• **Fruits** Tanzania[33, 34, 36, 47, 48] Kenya[33, 40] Uganda[43, 44, 55, 57, 58] Zanzibar Island & Pemba Island[53, 56]
• **Cereals** Tanzania[47] Kenya[47] Uganda[47]	• **Spices** Tanzania[33, 34, 59-61] Kenya[33, 60-62] Uganda[44, 55, 61] Zanzibar Island & Pemba Island[53, 56]
• **Legumes** Tanzania[23, 26, 33, 34, 36, 48] Kenya[40, 41, 49] Uganda[50]	• **Oils and fats** Tanzania[47, 63] Kenya[47] Uganda[44, 47, 55, 64] Zanzibar Island & Pemba Island[47, 53, 56]
• **Root and tubers** Tanzania[26, 34, 36, 47, 51-53] Kenya[40, 49] Uganda[47]	• **Traditional drinks** Tanzania[33, 34 62, 48, 65 48, 66] Kenya[39, 40, 49, 54] Uganda[45, 46, 55, 67, 68]
• **Vegetables** Tanzania[19, 31, 33, 34, 36, 37, 48, 52] Kenya[40, 54] Uganda[55] Zanzibar Island & Pemba Island[53, 56]	• **Animal foods** Tanzania[34, 26, 31-33, 36, 37, 48, 52, 59, 66] Kenya[40] Uganda[43, 55] Zanzibar Island & Pemba Island[53, 56]
II. Cultural food habits	
• **Diet and dishes** Tanzania[31-33, 35, 34, 36, 38, 65, 69] Kenya[41, 54, 62, 70, 39, 40, 49, 71] Uganda[43-46, 50, 55, 64, 68, 72, 73] Zanzibar Island & Pemba Island[53, 56]	• **Children** Tazania[32, 66, 69] Kenya[39-41, 54, 62, 70, 79] Uganda[32, 44, 58, 64, 67, 68]
• **Taboos and ritual foods** Tanzania[33, 74-76] Kenya[74, 76] Uganda[44, 55, 64, 74, 77]	• **Agriculture** Tanzania[34, 38, 65, 69, 80, 81] Kenya[30, 39, 78] Uganda[20, 44, 55, 68, 82]
• **Cooking methods and preparation** Tanzania[33, 34, 36, 38, 48, 69, 78] Kenya[39, 40, 49, 54] Uganda[44, 50, 73] Zanzibar Island & Pemba Island[53, 56]	• **Local markets** Tanzania[63] Kenya[49]
• **Women** Tanzania[32, 36, 66, 69, 75] Uganda[32, 44, 57, 58, 64, 67, 68]	
III. Nutrients and health	
• **Calories** Tanzania[33-35, 37, 38, 69, 83-85] Kenya[27, 30, 39, 86] Uganda[64, 73, 77, 87] Zanzibar Island & Pemba Island[56]	• **Vitamins and Minerals** Tanzania[33-35, 37, 38, 69, 83-85] Kenya[27, 30, 39, 86] Uganda[64, 73, 77, 87]
• **Protein** Tazania[33-35, 37, 38, 69, 83-85] Kenya[27, 30, 39, 86] Uganda[64, 73, 77, 87] Zanzibar Island & Pemba Island[56]	• **Health and disease** Tanzania[38, 52, 65, 69, 83, 85, 88-90 13, 19, 34, 80] Kenya[30, 39, 86, 91-95] Uganda[45, 46, 55, 64, 68, 96]

Cereals

The main cereal staples of East Africa were millet and sorghum. They were an important energy source, and in certain seasons of the year they supplied 80 to 90% of the dietary protein intake, and virtually all the vitamin B_1, nicotinic acid, vitamin A, calcium and phosphorus intake. Millet is one of the oldest grains and possibly the first used as a staple food. It is believed to have originated in Uganda, a region considered by some to have been the *Breadbasket of Africa*.[41] Millet is known for its high calcium content. It can grow in poor soil and mature quickly if adequately irrigated.[42] Sorghum bicolor was reportedly drought resistant.[42] These cereals are high in calcium, carotene and protein.[42]

Legumes

Kidney beans, cowpeas (*Vigna unguiculata*), groundnuts (*Arachis hypogaea*) and bambara groundnuts (*Vigna subterranea*) were commonly consumed throughout Tanzania.[23, 26, 32, 40, 43, 44] In Kenya, legumes consumed among the *Kikuyu* included several bean and pea varieties, including a European dwarf bean called *mboco*, pigeon pea (*njugu*), black bean, brown or white bean varieties (*njahɨ*, red bean (*kunde*), small peas (*thoroko* or *chirok*), small green pea (*thuu*), and small round green pea (*podzo*).[45-47] In Uganda, approximately one sixth of the total 6.3 million cultivated acres were occupied by grain legumes, of which groundnuts and kidney beans were the most prevalent, followed by cow peas, pigeon peas, and field peas.[48]

Root and tubers

The most common root and tubers cultivated in East Africa during this period included tannia (*Xanthosoma sagittifolium*), taro (*Colacasia esculenta*) and various yams (*Dioscorea spp.*). More recently introduced roots and tubers include cassava and sweet potato.[26, 32, 38, 40, 42, 49]

Vegetables

Thirty-nine wild, indigenous and edible vegetables were identified in Tanzania including twenty-one on Ukara Island[32], three in the Tabora region[40] and fifteen among the Sukuma in the Lake Province.[43]

Common traditional leafy vegetables that formed an essential part of the East African diet included amaranth (*Gynandropsis gynandra*), baobab (*Adansonia digitata*), African nightshade (*Solanum spp.*), hibiscus (*Hibiscus sabdariffa*), spiderplant (*Basella alba*) and taro leaves (*Colacasia esculenta*).[19, 32, 34, 40, 43, 44, 50, 51] Leaves of introduced vegetables such as sweet potato, pumpkin, cassava, beans and cowpeas were also consumed when in season.[38]

Fruits

Among the *Bahaya*, who occupied the northwestern corner of Tanzania, locally grown fruits included oranges, tangerines, lemons, limes, pineapples, pawpaw, passion fruit, mangoes, tree tomatoes, sweet and yellow bananas, plantain and bitterberries.[44] Sixteen wild and edible fruits of the *Sukuma* in the Lake Province have also been documented.[43] The pawpaw and cape gooseberry were a favorite snack for women and children in the Kiberege division of the Ulanga Valley in Tanzania.[42,50] On Ukara Island, fruits including unripe lemons were eaten by children.[52] Wild fruits on this island included *mfiru, sungwa, ndobe, mamonyi, mande, buyeko,* and *buhunda*.[52] In Uganda, figs, tamarinds, shea butter fruits and the fleshly part of borassus palm fruits were commonly assimilated into the regular diet.[53] Among the *Baganda*, the largest ethnic group in Uganda during the 1960s, pawpaw and passion fruit were the most common fruits.[31] *Ntula*, berries of *Solanum sp.*, were considered a snack food for children.[31]

Spices

Spices were an integral part of the food culture in coastal regions due to longstanding Islamic and Indian influence. The most common varieties used were black pepper, chili pepper, *capsicum annum* (*Pilipili kali*), cinnamon, curry powder (*Bizari*), and tamarind.[54, 55] Seven surveys reported that use of salt was widespread in East Africa.[31, 32, 34, 43, 56-58] Reeds were burnt and salt was obtained from the ashes by a process of solution and reprecipitations.[43] In Teso, Uganda, salt was obtained by dissolving it from the ashes of *echuga* (*Leonolis meptifolia*), *epungula* (*Coreopsis ugandensis*), *elokile* (*Sonchus bipontini*), *epopong* (*Euphorbia candelubra*), *essege* (*Pennisetum sp.*) and *eliloto* (*Sesamum macranthum*).[31] Besides salt, the ripe fruits of *elamai* (*Ximenia americana*) and the pods of *epiduru* (*Tamarindus indica*) were used to flavor the foods.[34]

Oils and fats

The main sources of vegetable fats used in East Africa included the oil of simsim (*Sesamum indicum*), cottonseed, shea butter nut (*Butyrospermum parkii*), coconut, groundnut and palm fruit.[31, 34, 42] Oyster nuts (*Telfairia pedata*) called *kweme* in Kiswahili or *nkungu* in Kisambaa, were highly esteemed for pregnant women among several ethnic groups in East Africa.[59] This particular nut was believed to promote lactation due to a high content of protein and fat.[59]

Traditional drinks

Eleven reports[34, 44, 60-62 32, 36, 43, 46, 63, 64] described the use, preparation and importance of native beer in East Africa. Beer was prepared from a variety of constituents, including millet, sorghum or maize, and sugar cane or honey.[34, 44, 60-62 32, 36, 43, 46, 63, 64] The preparation and use of other native drinks were reported in five publications including the descriptions of a raw defibrinated blood and milk drink in Karamoja, Uganda,[65] banana wine in the Bukoba district of Tanzania,[44] *usawo*, a mixture of cow's blood, sour milk and honey consumed by the *Chagga* mothers of northern Tanzania, after delivery,[66] and *ucuru* a thin gruel prepared from finger millet or maize and consumed by the *Kikuyu* in Kenya.[45, 64]

Animal foods

Most groups in East Africa ate meat only occasionally.[53] Where fish was available, generally only around the lake regions and the coast, it was consumed both fresh and dried.[53] Various species of birds, rats, mice, locusts, grass hoppers and white ants, generally relished as delicacies, were also consumed.[53] Meat and milk were more significant among pastoral people such as the *Maasai Samburu* and *Turkana*, who live in the dry steppes.[17] Milk was obtained from cows, goats and occasionally sheep. It was taken fresh or was fermented in containers, mainly gourds (*kibuyu* in Swahili) or hollowed-out wood, as in the case of many pastoralists in the north and east.[13] The milk was churned to make sour milk or butter, popular among pastoralists.[13, 53, 57, 67]

Food preparation and culture

Examples of traditional dishes and meal patterns in Kenya, Tanzania and Uganda are presented in Table 4-6.

Taboos and ritual foods

Food taboos most commonly existed among women. These could include the avoidance of eggs, chicken, mutton and several species of fish.[68, 69] Clear indications of why these foods were ritually avoided were not always collected by the researchers.

Cooking methods and preparation

Unique utensils used for food preparation in East Africa have been documented in several publications.[64, 32, 48, 70] Most foods were boiled, and occasionally roasted.[36] Food preparation techniques have been described for cereals, roots and tubers, legumes, vegetables and animal foods in Tanzania,[43, 50, 67, 71, 32, 37, 40, 63, 70] Kenya[36, 45, 46, 64] and Uganda.[31, 34, 35, 48, 53, 61, 62, 65, 72, 73]

Table 4. Main dishes of different ethnic groups in Kenya.

Author[Ref.] (year)	Location (Ethnical group)	Main dishes
Callanan[36] (1926)	Nyanza Province (*Luo*)	• *Kuon* (*ugali* in Swahili): A doughy substance prepared by boiling *mtama* (sorghum flour) or flour of finger millet (*kal* in Swahili) in water until a doughy substance was formed - eaten with meat or native vegetables, buttermilk (*buyo*), blood, fish, chicken or eggs • Wimbi (Finger millet - *Eleusine coracana*), was chiefly consumed by the *Luo* near the Kisii border • Less frequent: A mixture of beans (*oganda*) and maize (*oduma*) termed *nyoyo*

Orr et al.[67] (1931) Farnworth[44] (1937)	Central Province (*Kikuyu*)	• *Irio* A mixture composed of maize, various kinds of beans and mashed bananas
		• Gruel (*ucuru*) made from millet flour and water (or other beverage)
		• Women's dishes were distinguished by containing:
		  a. Green leaves
		  b. Special millet varieties (e.g. red millet varieties: *mugimbi* or *mwimbe*)
		  c. Salt or salt substitutes

Allen[46] (1955)	Costal region (*Giriama*)	• *Sima* A porridge made from maize and *tui*[*] • Porridge made from brown rice, cassava and banana • *Kitowea*: A mixture made from *kunde* (read bean), *podzo* (small, round green bean) stewed beef or goat, boiled fish or shark, prawns or chicken • Bananas - boiled, eaten raw, or fried in ghee • Sweet potatoes roasted in ashes or boiled
Shaper et al.[68] (1961)	Northern Kenya (*Samburu*)	• Staple food: milk - warriors usually drank milk twice a day • Other major dietary item: meat • Meat and milk were never consumed on the same day • Supplementary food: blood - only used during the dry season
Gerlach[69] (1961)	Kenyan coast strip (*Digo*)	• *Breakfast:* cold remains of supper or thin gruel, tea and raised cake - made of maize meal or imported wheat flour • *Midday*: same foods as for breakfast or roasted corn, bean soup or roasted, baked bananas • *Evening:* Digo (>8 years old) consumed one quart of a thick mixture of rice, plantain, sweet potatoes (*chakuria*) and 1 half pint of meat, fish, sour milk or wild greens (*chitoweo*) excluding beans and bananas

[*]Extract obtained through washing and squeezing grated coconuts.

Table 5. Main dishes of different ethnic groups in Tanzania.

Author[Ref.] (year)	Location (Ethnical group)	Main dishes
Culwick et al.[49] (1939)	Ulanga, Kiberege Division	• Boiled rice or porridge (made from maize, cassava, finger millet) with a relish of fish or meat and wild green vegetables
Laurie et al.[32] (1951)	Bukoba district (*Bahaya*)	• Plantains, beans and *ningu* (*Labeo victorianus*) or other fish
McLaren[62] (1960) Tanner[42] (1956)	North-West Tanzania (*Sukuma*)	• 2 main meals/day - at noon and around sunset • Stiff porridge made from sorghum, millet, maize flour or cassava - eaten together with a relish of green vegetables, a meat or fish (dried) stew • Maize on the cob, cassava, ground nuts, tomatoes and other fruits were eaten raw
Schaffer et al.[37] (1963)	Central Province (*Gogo*)	• *Ugali** was made from millet, maize or sorghum - eaten with at least one sauce (made from soured milk, legumes, leafy vegetables or meat) • Most favourite sauce was prepared from sesame, cowpea leaves, tomatoes, onions and aloes • Meat was on average consumed once a week • Soured milk was served with 50% of the meals • Sweet potatoes or pumpkins were served as *soul food*
Jelliffe et al.[66] (1962)	North Tanzania (*Hadza*)	• Food was eaten after it was obtained through hunting or gathering • Meat and yams were barbecued - only older women were permitted to have cooking pots to boil the meat in • Wild fruit, berries and seeds were eaten raw

*Mixing flour from a starchy food in hot water and cooking as one mixes the substance to a paste that varied in consistency
**Porridge based on cereals, its consistency varied among the areas as well as the flavoring (salt, sugar, lemon, tamarind, baobab, coconut, cow ghee/butter or milk)
#Extract obtained through washing and squeezing grated coconuts
+Zanzibar and Pemba Island were inhabited by 5 ethnic groups

Table 5 continue. Main dishes of different ethnic groups in Tanzania.

Author[Ref.] (year)	Location (Ethnical group)	Main dishes
Nguma[70] (1969)	Kilosa district (*Wasagara, Wakaguru, Wavidunda, Waluguru*)	• 1-3 meals/day • *Ugali** was made from maize and sorghum flour and was consumed with beans, cowpeas, pigeon, peas and meat, fish or green leaves • Snacks: banana, cassava and corn
Tanzania National Nutrition Unit[40] (1967)	Tabora region (*Nyamwezi*)	• *Breakfast:* tea or coffee with sugar and sometimes milk • *Midday/Evening: ugali** was made from cassava, maize or sorghum flour and eaten with a side dish prepared from green leaves or legumes • *Foods for special occasions:* rice served with meat or chicken as a side dish • *Foods consumed during work:* boiled cassava, tea, groundnuts, *makande* (maize and bean mixture), *ugali**, vegetables, *uji***, fresh or dried cassava

Zanzibar Protectorate[54] (1937) Smith et al.[53] (1935)	Zanzibar and Pemba Island[+]	• *Breakfast:* tea - if affordable, small amount of tinned or fresh milk with sugar added, white bread or otherwise *makake wa kusukuma* (gruel of millet), fresh fruits or dates
		• *Midday:*
		  a. Fish with one of the following foods: cassava, plantain, sweet potato, yam, breadfruit
		  b. *Mseto* (i.e. rice and *dhal* boiled together with *tui*[#])
		  c. *Kiwanda* (eggs beaten up and fried - eaten with rice and raw green leaves)
		• *Evening:* fish or meat curry with rice or bread - main meal of the day
		• Boiled green leaves were sometimes eaten as a 3rd dish
		• *For a feast: pilau* - a dish made with beef, mutton or goat's meat, ghee, gamti rice and bread

*Mixing flour from a starchy food in hot water and cooking as one mixes the substance to a paste that varied in consistency
**Porridge based on cereals, its consistency varied among the areas as well as the flavoring (salt, sugar, lemon, tamarind, baobab, coconut, cow ghee/butter or milk)
#Extract obtained through washing and squeezing grated coconuts
+Zanzibar and Pemba Island were inhabited by 5 ethnic groups

Table 6. Main dishes of different ethnic groups in Uganda.

Author[Ref.] (year)	Location (Ethnical group)	Main dishes
Courcy-Ireland et al [34] (1937)	Teso Ajuluku and Opami village (*Iteso*)	• 2 main meals/day (midday and evening) • *Breakfast*: drink of beer, a baked sweet potato or cassava - eaten in fields during work • Fish was consumed on a regular basis in Opami compared to Ajuluku • *Atap:* ground and cooked wimbi (finger millet) • Milk was used for *atap* instead of water by wealthier families and the curds were mixed with greens or other relishes
Rutishauser[31] (1963)	Buganda (*Baganda*)	• *Breakfast:* left-over food from the night before which was also given to children as a midday meal • *Midday/Evening:* one staple (i.e. *matoke* or other types of bananas, sweet potato, cassava, maize or yams) and one or more sauces made from gathered insects, fish, meat, beans, sesame, groundnuts or wild vegetables
Jelliffe et al.[35] (1963)	Acholi district (*Acholi*)	• Stiff dough-like preparation of finger millet (*kwon*) - eaten with *dek,* a variable mixture of: beans (*Phaseolus vulgaris*), cowpeas *(Vigna unguiculata)*, congo peas (*Cajanus indicus*), simsim (*Sesami indicum*) and meat or fish • When available, mushrooms, termites, wild edible leaves, honey and fruits such as mangoes were eaten

Women and infants

In Tanzania, modifications of the diet in pregnant and lactating women, as well as the diets of infants were investigated in four studies.[37, 40, 66, 67] Moller[74] reported on the different customs and beliefs involved in pregnancy and delivery of newborns or twins among several ethnic groups in Tanzania, including the *Wahehe, Wagogo, Waluguru, Sukuma, Wanyakyusa, Wachaga,* and *Bahaya.*

In Uganda, seven publications have provided an insight into the cultural and anthropological factors in maternal nutrition, lactation and child feeding practices among the Baganda,[68] Buganda,[31] Luo,[75] Acholi,[35] Hadza,[67] Karamojong,[61] and *Bahaya*.[60]

Children

In Tanzania, the feeding of children has been described among the *Hadza* hunters,[67] *Gogo*,[37] and *Chagga*.[66] In Kenya, the feeding of children has been described amongst the *Luo*,[36, 64] *Kikuyu*,[45, 47, 64, 76] *Maasai*[77] and *Samburu*.[78] In Uganda, such descriptions have been collected amongst the *Buganda*,[31] *Baganda* and *Luo*,[75] *Acholi*,[35] *Hadza*,[67] *Karamojong*[61] and *Bahaya*.[60]

Agriculture

Agricultural practices have been documented in Tanzania among the *Wangoni*,[79] *Bahaya*,[32] *Gogo*,[37, 63] *Sukuma*,[63] *Wachaga*,[70] *Shambala*,[80] in Kenya among the *Kikuyu*,[30, 64] *Luo*,[64] *Wakamba*,[64] *Digo* and *Daruma*[81] and in Uganda
among several unnamed tribes.[20, 31, 34, 61, 82]

Local markets

Descriptions of traditional foods offered at local markets, including seasonal variations, supply and demand, and variations in prices have been provided for Bumbuli, Tanzania[59] and the *Kikuyu* market in the Central Province of Kenya.[46]

Dietary intake and health status indicators

Nutritional and health status, as well as dietary intake was evaluated in 19 publications.[16]

2.4. Discussion

The principale investigator (V.R.) systematically reviewed the *Oltersdorf collection*, 75 observational studies conducted between the 1930s and 1960s and collected by the *Max-Planck Nutrition Research Unit*, formerly located in Bumbuli, Tanzania. The primary intention was to extract data pertaining to traditional East African foods and food habits. This data extraction revealed several important findings which may have profound implications for averting current NCD trends and malnutrition throughout sub-Saharan Africa.

The earliest food crops used by most agriculturalists in East Africa included sorghum, finger and pearl millets, hyacinth (*lablab*) beans, bambara groundnuts, bottle gourds, cowpeas, and yams.[83] According to our data extraction, several of these food crops remained part of the diet in various parts of East Africa between the 1930s and 1960s. Many have recently been validated for their significant health benefits. For example, sorghum has recently been identified as containing significant amounts of polyphenols and antioxidants,[84] while millet has been shown to reduce risk factors for CVD.[85, 86] According to Mossanda *et al.*,[87] the bambara groundnut (*Vignea subterranea*) possesses anti-oxidative and anti-mutagenic activities.

Uncultivated and wild edible fruits, vegetables and other plants species could supply significant amounts of micronutrients to the diet of the Africans.[88] Green leafy vegetables are high in beta-carotene,[89] contain significant amounts of polyphenols and have free radical scavenging abilities.[90] The *Luo* people of Western Kenya have suggested that the leafy vegetables that form an important part of their diet protect against gastro-intestinal disturbances. In particular, *Solanum nigrum* is effective against the protozoan gut parasite *Giardia lamblia*.[91] Okra (*Hibiscus esculentus*) has been identified as a cholesterol lowering food.[92] Its ability to reduce total cholesterol and low density lipoprotein cholesterol may contribute to the prevention of CVD.

Edible wild roots and tuber species have been reported to be an important energy and water source for pastoralists and hunter-gatherers and were well-recognized for their medicinal properties by several ethnic groups throughout East Africa.[67, 78, 93, 94] A recent investigation by Hou et al.[95] has revealed that the storage protein of yam tuber (*Dioscorea batatas*) may have antioxidant properties.

Traditionally, plant oils were used for cooking in East Africa. These included the oil of shea butter nut (*Butyrospermum parkii*), sesame seed (*Sesamum indicum*) and red palm. These particular plant oils are associated with nutrition and health benefits.[96-99] Sesame oil, for example, improves hypertension, lipid profiles, and lipid peroxidation and increases enzymatic and nonenzymatic antioxidants.[96] Owing to its high content of phytonutrients and antioxidant properties, the possibility exists that palm fruit offers health benefits by reducing lipid oxidation, oxidative stress and free radical damage.[97, 98] The use of palm fruit or its phytonutrient-rich fractions, particularly water-soluble antioxidants, may confer some protection against a number of disorders and diseases including CVD, cancers, cataracts and macular degeneration, cognitive impairments, and Alzheimer's disease.[97, 98]

Animal foods played a significant role in certain East African communities throughout this period. Although the diet might have been high in animal foods, the evidence suggests an absence of NCDs and a low CVD risk profile.[13, 100] A combination of factors, including the lack of processed meats, low energy intake, and deep-rooted cultural practices which included ample amounts of physical activity may have been essential in averting CVD. By contrast, the processed meats saturating markets today have been directly associated with an elevated CVD risk profile.[101] Recent evidence suggests that the regular chewing of *Commiphora* and other species of *myrrh sensu lato*, high in hypoglycemic,[102] antioxidant,[103] anti-inflammatory, and antibiotic properties[104] offers health benefits. The antioxidant activity of phenolics[105] and the cholesterol-binding activity of saponins[100] in *Maasai* food may further modulate the effects of a high-lipid diet.[106]

Between the 1930s and 1960s, most foods in East Africa were boiled or roasted, whereas many foods today are fried or deep fried.[36, 107] Frying and deep frying in particular, have been associated with adverse health outcomes including elevated concentrations of pro-inflammatory cytokines and homocysteine.[101]

A diversified diet combined with traditional knowledge of food preparation may advance our understanding of health by complementing current scientific knowledge of micronutrient density and nutrient absorption.[108] A diet consisting of a diversity of wild, edible plants, legumes, condiments, wild game meat, milk, fish and cereals such as sorghum and different varieties of millet is likely to be associated with significant health benefits and longevity.[109-111] In Kenya, Onyango and colleagues have recently demonstrated that a diversified diet enhances the development of young children.[112]

Governmental (i.e. corporate) policies and their major socio-economic and environmental repercussions have fundamentally destroyed human health in Africa.[113] In addition, the simplification of a now *globalized* diet has presented the global population with unprecedented obstacles. Health implications of dietary simplification include nutrient deficiency and an increased prevalence of NCDs.[114] These trends are occurring worldwide, in both developed and developing countries, including the countries of East Africa.

Knowledge of traditional food habits in Africa is being lost. There is clearly an imperative need for documentation, compilation, and dissemination of this rapidly eroding wealth of information. Knowledge of traditional African food habits could be used to improve the lives of African residents and African migrants, including the recent surge of African migrants entering Australia and New Zealand. Further, such information can and should be utilized by the global community for improving the current *globalized* food culture, which has largely been responsible for the obesity and diabetes epidemics currently plaguing the world. The online collection,[16] could be instrumental in disseminating information related to traditional African foods and food habits.

In summary, it appears that potential did exist for a rich food culture in East Africa from the 1930s to the 1960s, despite years of imperial occupation. Many of the traditional foods presented in this review have known health benefits according to the latest annals of scientific inquiry. The knowledge which has evolved and remains at the heart of this *Cradle of Civilization*, including incredible knowledge of the relationship between food, food habits and longevity, should not be ignored and should indeed be investigated further. Such inquiry may be required for the human species to transcend current NCDs epidemics, and move toward a paradigm based upon holistic health and well-being.

2.5. Résumé

Background: Knowledge of traditional African foods and food habits has been, and continues to be, systematically extirpated. With the primary intent of collating data for our online collection documenting traditional African foods and food habits (available at: www.healthyeatingclub.com/Africa/), the *Oltersdorf Collection* was reviewed, 75 observational investigations conducted throughout East Africa (i.e. Tanzania, Zanzibar Island and Pemba Island, Kenya, and

Uganda) between the 1930s and 1960s as compiled by the *Max-Planck Nutrition Research Unit*, formerly located in Bumbuli, Tanzania. Methods: Data were categorized as follows: (1) food availability, (2) chemical composition, (3) staple foods (i.e. native crops, cereals, legumes, roots and tubers, vegetables, fruits, spices, oils and fats, beverages, and animal foods), (4) food preparation and culture, and (5) nutrient intake and health status indicators. Results: Many of the traditional foods identified, including millet, sorghum, various legumes, root and tubers, green leafy vegetables, plant oils and wild meats have known health benefits. Food preparatory practices during this period, including boiling and occasional roasting are superior to current practices which favor frying and deep-frying. Overall, our review and data extraction provide reason to believe that a diversified diet was possible for the people of East Africa during this period (1930s-1960s). Conclusions: There is a wealth of knowledge pertaining to traditional East African foods and food habits within the *Oltersdorf Collection*. These data are currently available via the online collection. Future efforts should contribute to collating and honing knowledge of traditional foods and food habits within this region, and indeed throughout the rest of Africa. Preserving and disseminating this knowledge may be crucial for abating projected trends for non-communicable diseases and malnutrition in Africa and abroad.

2.6. References

1. Zimmet P. Globalization, coca-colonization and the chronic disease epidemic: Can the Doomsday scenario be averted? *J Intern Med.* 2000; 247:301-310.
2. World Health Organization. Life in the 21st century: A vision for all. The world Health Report 1998. Geneva: World Health Organization;

1998:1-241.
3. Leeder S, Raymond S, Geenberg H, Liu H, Esson K. *Race Against Time: The Challenge of Cardiovascular Disease in Developing Economies.* New York: Columbia University; 2004.
4. Wild S, Roglic G, Green A, Sicree R, King H. Global prevalence of diabetes: Estimates for 2000 and projections for 2030. *Diabetes Care.* 2004; 27:1047-1053.
5. Yamori Y, Miura A, Taira K. Implications from and for food cultures for cardiovascular diseases: Japanese food, particularly Okinawan diets. *Asia Pac J Clin Nutr.* 2001; 10:144-145.
6. Cockerham W, Yamori Y. Okinawa: An exception to the social gradient of life expectancy in Japan. *Asia Pac J Clin Nutr.* 2001; 10:154-158.
7. Trichopoulou A, Critselis E. Mediterranean diet and longevity. *Eur J Cancer Prev.* 2004; 13:453-456.
8. Serra-Majem L, Roman B, Estruch R. Scientific evidence of interventions using the Mediterranean diet: A systematic review. *Nutr Rev.* 2006; 64:27-47.
9. MacLennan R, Zhang A. Cuisine: The concept and its health and nutrition implications - global.*Asia Pac J Clin Nutr .* 2004; 13:131-135.
10. Liu X, Li Y. Epidemiological and nutritional research on prevention of cardiovascular disease in China. *Br J Nutr.* 2000; 84:199-203.
11. Hermanussen M, Garcia A, Sunder M, Voigt M, Salazar V, Tresguerres J. Obesity, voracity, and short stature: The impact of glutamate on the regulation of appetite. *Eur J Clin Nutr.* 2006; 60:25-31.
12. Johns T, Nagarajan M, Parkipuny M, Jones P. Maasai gummivory:

Implications for paleolithic diets and contemporary health. *Curr Anthropol.* 2000; 41:453-459.

13. Mann GV, Schäffer RS, Anderson RS, Sandstead HH. Cardiovascular disease in the Masai. *J Atheroscler Res.* 1964; 4:289-312.

14. Oniang'o RK. Food habits in Kenya: The effects of change and attendant methodological problems. *Appetite* 1999; 32:93-96.

15. Kuhnlein H, Receveur O. Dietary change and traditional food systems of indigenous peoples. *Annu Rev Nutr.* 1996; 16:417-442.

16. Raschke V. East African food habits on-line. In: Wahlqvist ML. HEC Press. Available at: http://www. healthyeatingclub.org/Africa/

17. Oltersdorf U. Comparison of nutrient intakes in East Africa. Paper presented at: *The Human Biology of Environmental Change.* 5-12 April, 1971; Blantyre, Malawi. 1971:54-59.

18. Gale GW. Food balance sheets for the African populations of East Africa. *East Afr Med J* 1960; 37:410-417.

19. Latham M. Nutritional Studies in Tanzania, (Tanganyika). *J Am Diet Assoc* 1967; 7:31-71.

20. Cleave J. Food consumption in Uganda. *East Afr J Rural Dev.* 1968; 1(1):70-87.

21. Government of Uganda. Annual Report of the Medical Department for 1950. Entebbe: Government Printer; 1951.

22. Wilson W. *The adequacy of the food supply for Uganda Africans.* Memorandum of the Nutrition Advisory Committee of the Uganda Government (unpublished). 1954

23. French MH. Some notes on the common foodstuffs used in the diet of East African natives. *East Afr Med J.* 1937; 13:374-378.

24. Naik A. Chemical composition of Tanzania feedstuffs. *East Afr*

Agric For J. 1969; 32(2):201-205.
25. McLaren D. Sources of carotenoid and vitamin A in Lake Province, Tanganyika. *Acta Trop.* 1961; 18:78-80.
26. Schlage C. *Analysis of some Important Foodstuffs of Usambara.* In: Kraut H, Cremer HD, eds. Investigations into Health and Nutrition in East Africa, Afrikastudien. München: Weltforum Verlag, München. 1969; 42:85-178.
27. Orr J, Gilks J. The Masai and Kikuyu diet. Medical Research Council. Special Report Series No.155; 1951:22-62.
28. Muskat E, Keller W. Nutzung von Wildpflanzen als zusätzliche Vitaminquelle der Ernährung der Bevölkerung Kenias. *Internat Z Vit Forschung* 1968; 38:538-544.
29. Harvey D. *The chemical composition of some Kenya foodstuffs.* Records of the Medical Research Laboratory No. 10. Colony and Protectorate of Kenya 1951:1-10.
30. Keller W, Muskat E, Vlader E. *Some Observations Regarding Economy, Diet and Nutritional Status of Kikuyu Farmers in Kenya.* In: Kraut H, Cremer, eds. Investigations into Health and Nutrition in East Africa, Afrikastudien. München: Weltforum Verlag, München. 1969; 42:241-266.
31. Rutishauser I. Custom and child Health in Buganda. *Trop Geogr Med.* 1963; 15:138-147.
32. Laurie W, Brass W, Sewell P, et al. *A health survey in Bukoba district, Tanganyika.* East African High Commission; 1951:1-138.
33. Korte R. *The Nutritional and Health Status of the People Living on the Mwea-Tebere Irrigation Settlement.* In: Kraut H, Cremer HD, eds. Investigations into health and nutrition in East Africa. München: Weltforum Verlag München 1969; 42:267-331.

34. Courcy-Ireland M, Hosking H, Loewenthal L. An investigation into health and agriculture in Teso, Uganda. Teso: Agricultural Survey Committee; Nutrition Report No. 1; 1937:1-28.
35. Jelliffe DB, Bennett JF, Stroud CE, Welbourn HF, Williams MC, Jelliffe EFP. The health of Acholi children. Community and child health studies in East Africa No. 5. *Trop Geogr Med.* 1963; 15:411-421.
36. Callanan J. Notes on the foodstuffs of the Luo tribe. *Ken Med J.* 1926; 3:58-60.
37. Schaffer R, Finklestein F. The food and growth of Gogo children. Stencilled Paper 1963:1-10.
38. Kreysler J, Schlage C. *The Nutrition Situation in the Pangani Basin.* In: Kraut H, Cremer HD, eds. Investigations into health and nutrition in East Africa. München: Weltforum Verlag München 1969; 42:85-178.
39. Kasper H, Schaefer HJ, Theermann P. *A Contribution to Determine the Nutritional State of the People living in the Central Province of Kenya.* In: Kraut H, Cremer HD, eds. Investigations into health and nutrition in East Africa. München: Weltforum Verlag München 1969; 42:227-240.
40. Tanzania National Nutrition Unit. Report of a dietary survey in the Tabora region. Tanzania: The Tanzania National Nutrition Unit; Tanzania Nutrition Committee Report Series 4a; 1967:1-11.
41. Whitney N. A talk radio first: Nicole interviews both David Icke & Zulu Shaman/South African elder Credo Mutwa. CFUN 1410. June 25, 2005.
42. Culwick A, Culwick G. Nutrition and native agriculture in East Africa. *East Afr Med J.* 1941; 6:146-153.

43. Tanner R. A preliminary enquiry into Sukuma diet in the Lake Province, Tanganyika territory. *East Afr Med J.* 1956; 33:305-324.
44. Laurie W, Trant H. A health survey in Bukoba district, Tanganyika, East African Medical Survey Monograph No. 2. Nairobi: East African High Commission; 1954:1-32.
45. Farnworth A. Kikuyu diet. *East Afr Med J.* 1937; 14:120-131.
46. Procter R. The Kikuyu market and Kikuyu diet. *Ken Med J.* 1926; 3:15-22.
47. Allen KW. The monotonous diet of the African. *East Afr Med J.* 1955; 32:95-97.
48. Dean R. Protein supply in Uganda. *East Afr Med J.* 1962; 39:493-500.
49. Raymond W, Jojo W, Nicodemus Z. The nutritive value of some Tanganyika foods II - Cassava. *East Afr Med J.* 1941; 6:154-159.
50. Culwick A, Culwick G. A study of factors governing the food supply in Ulanga, Tanganyika Territory. *East Afr Med J.* 1939; 16:43-61.
51. Popleau W, Schlage C, Keller W. *Nutrition and Health in Usambara.* In: Kraut H, Cremer HD, eds. Investigations into health and nutrition in East Africa. München: Weltforum Verlag München 1969; 42:25-45.
52. Brass M, Laurie H, Trant H. Ukara Survey: East African High Commission; 1951:27-55.
53. Uganda Protectorate Nutrition Committee. Review of nutrition in Uganda. Entebbe: Nutrition Committee, Government Printer Uganda; 1945:1-19.
54. Smith W, Smith E. Native diet of Zanzibar. *East Afr Med J.* 1935; 11:246-251.
55. Zanzibar Protectorate. Nutritional review of the natives of Zanzibar.

Zanzibar Protectorate; 1937:1-20.
56. Foster J. "Pica". *East Afr Med J.* 1927:9.
57. Jerrard R. Tribes of Tanganyika, their districts, usual dietary and pursuits. Dar es Salaam: Government Printer; 1935:1-24.
58. Worthington E. On the food and nutrition of African natives. *Africa.* 1936; 9:150-165.
59. Schlage C. *Observation at the Local Market in Usambara.* In: Kraut H, Cremer HD, eds. Investigations into health and nutrition in East Africa. München: Weltforum Verlag München 1969; 42:72-84.
60. Rwegelera. Tribal custom in infant feeding among the Haya. *East Afr Med J.* 1963; 40:366-369.
61. Jelliffe D, Jelliffe E, Benett F, White R. Ecology of childhood disease in the Karamojong in Uganda. *Arch Environ Health* 1964; 9:25-36.
62. Loewenthal J. An inquiry into vitamin A deficiency among the population of Teso, Uganda, with special reference to school children. *Ann Trop Med.* 1935; 29:349-360.
63. McLaren D. Nutrition and eye disease in East Africa: Experience in Lake and Central Provinces, Tanganyika. *J Trop Med Hyg.* 1960; 63:101-122.
64. Bohdal M, Gibbs N, Simmons W. Nutrition surveys and campaigns against malnutrition in Kenya 19641968 . Geneva: WHO/FAO/UNICEF; 1969:1-171.
65. Holmes E, Stanier M, Thompso M. The serum protein pattern of Africans in Uganda: Relation to diet and malaria. *Trans Roy Soc Trop Med Hyg.* 1955; 49:376-384.
66. Lema N. Tribal customs in infant feeding II-Among the Chagga. *East Afr Med J.* 1963; 40:370-375.

67. Jelliffe D, Woodburn J, Bennett J. The children of the Hadza hunters. *J Pediatr* 1962; 60:907-913.
68. Jelliffe D, Bennett J. Cultural and anthropological factors in infant and maternal nutrition. *Fed Proc.* 1961; 20:185-187.
69. Kenneth M. Report on dietary surveys and nutritional assessment in East Africa. Nairobi: East African Bureau of Research in Medicine and Hygiene; 1953:1-18.
70. East African High Commission. Nutrition survey in the Kilimanjaro area: Government Printer; 1968:1-14.
71. Nguma G. Food Science and Applied Human Nutrition Research: Progress Report of a dietary survey in Kilosa district: Project No. 27; Research and Training Institute Ilonga Food Science Section. Tanzania Nutrition Unit Committee Report Series; 1969:1-11.
72. Benett F, Jelliffe E, Moffat M. The nutrition and disease pattern of children in a refugee settlement. *East Afr Med J.* 1968; 39:449.
73. Welbourn HF. The danger period during weaning (Part II). *J Trop Pediatr.* 1955; 1:98-105.
74. Moller M. African child health: custom, pregnancy and child rearing in Tanganyika. *J Trop Pediatr.* 1961; 7:66-80.
75. Welbourn HF. Notes on differences between Baganda and Luo children in Kampala. *East Afr Med J.* 1955; 32:291-298.
76. Orr J, Gilks J. *The Physique and Health of Two African Tribes.* London: His Majesty's Stationery Office; 1931:1-82.
77. Gower R. The effect of a change of diet on Masai schoolboys. Notes of the investigation into the health of Masai schoolboys at Monduli School during 1945. 1948:17-19.
78. Shaper A, Spencer P. Physical activity and dietary patterns in the Samburu of northern Kenya. *Trop Geogr Med.* 1961; 13:237-281.

79. Robson J. Malnutrition in Tanganyika. Tanganyika: Notes and Records. 1962; 58 59: 259-267.
80. Attems M, Keller W, Kraut H, Kreysler J, Popleau W, Schlage C. *Investigations in North East Tanzania*. In: Kraut H, Cremer, eds. Investigations into Health and Nutrition in East Africa, Afrikastudien. München: Weltforum Verlag, München 1969; 42:13-179.
81. Gerlach L. Socio-cultural factors affecting the diet of the northeast coastal Bantu. *J Am Diet Assoc.* 1964; 45:420-424.
82. Burgess AP. Calories and proteins available from local sources for Uganda Africans in 1958 and 1959. *East Afr Med J.* 1962; 39:449-465.
83. Maundu P, Ngugi G, Kabuye C. *Traditional Food Plants of Kenya*. Nairobi: Kenrik; 1999:1-269.
84. Dicko M, Hilhorst R, Gruppen H, et al. Comparison of content in phenolic compounds, polyphenol oxidase, and peroxidase in grains of fifty sorghum varieties from Burkina Faso. *J Agric Food Chem.* 2002; 50(13):3780 -3788.
85. Nishizawa N, Shimanuki S, Fujihashi H, Watanabe H, Fudamoto Y, Nagasawa T. Proso millet protein elevates high plasma level of high-density lipoprotein: A new food function of proso millet. *Biomed Environ Sci.* 1996; 9:209-212.
86. Gooneratne J, Munasinghe L, Senevirathne W. Millet bran and corn bran lowers plasma total and LDL cholesterol levels in hypercholesterimic subjects. *Ann Nutr Metab.* 2005; 49:304.
87. Mossanda K, Kundu J, Na HK, Beuth J, Chung AS, Joubert HF, Surh YS. Investigating African diet and beverage in Africans with low risk of digestive and liver cancers. *Ann Nutr Metab.* 2005;

49:168.

88. Grivetti L, Ogle B. Value of traditional foods in meeting macro - and micronutrient needs: the wild plant connection. *Nutr Res Rev.* 2000; 13:31-46.

89. Mulokozi M, Hedren E, Svanberg U. In vitro accessibility and intake of beta-carotene from cooked green leafy vegetables and their estimated contribution to vitamin A requirements. *Plant Foods Hum Nutr.* 2004; 59:1-9.

90. Oboh G, Akindahunsi A. Change in the ascorbic acid, total phenol and antioxidant activity of sun-dried commonly consumed green leafy vegetables in Nigeria. *Nutr Health.* 2004; 18:29-36.

91. Johns T, Faubert G, Kokwaro J, Mahunnah R, Kimanani E. Anti-giardial activity of gastrointestinal remedies of the Luo of *East Africa. J Ethnopharmacol.* 1995; 46:1-23.

92. Bangana A, Dossou N, Wade S, Guiro A, Lemonnier D. Cholesterol lowering effects of Okra (*Hibiscus esculentus*) in Senegalese adult men. *Ann Nutr Metab.* 2005; 49:199.

93. Shaper A, Jones M, Kyobe J. Plasma-lipids in an African tribe living on a diet of milk and meat. *Lancet.* 1961; 162:1324-1327.

94. Maundu P, Johns T, Eyzaguirre P, Smith F. Traditional root and tuber food plants of Sub Sahara Africa: Diversity and potential for improving health, nutrition and livelihood. Paper presented at: Cape to Cairo Safari Conference 14-19 September 2005, Potchefstroom, South Africa 2005:13.

95. Hou W, Lee M, Chen H, Liang WL, Han CH, Liu YW, Lin YH. Antioxidant activities of dioscorin, the storage protein of yam (Dioscorea batatas Decne) tuber. *J Agric Food Chem.* 2001; 49:4956-4960.

96. Sankar D, Sambandam G, Ramakrishna Rao M, Pugalendi K. Modulation of blood pressure, lipid profiles and redox status in hypertensive patients taking different edible oils. *Clin Chim Acta.* 2005; 355:97-104.
97. Sundram K, Sambanthamurth R, Tan Y. Palm fruit chemistry and nutrition. *Asia Pac J Clin Nutr.* 2003; 1:355-362.
98. Wattanapenpaiboon N, Wahlqvist M. Phytonutrient deficiency: the place of palm fruit. *Asia Pac J Clin Nutr.* 2003; 12:363-368.
99. Solomons N, Orozco M. Alleviation of vitamin A deficiency with palm fruit and its products. *Asia Pac J Clin Nutr.* 2003; 12:373-338.
100. Chapman L, Johns T, Mahunnah R. Saponin-like in vitro characteristics of extracts from selected non-nutrient wild plant food additives used by Maasai in milk and meat based soups. *Ecol Food Nutr.* 1997; 36:1-22.
101. Nettleton J, Steffen L, Mayer-Davis E, Jenny NS, Jiang R, Herrington DM, Jacobs DR Jr. Dietary patterns are associated with biochemical markers of inflammation and endothelial activation in the Multi-Ethnic Study of Atherosclerosis (MESA). *Am J Clin Nutr.* 2006; 83:1369-1379.
102. Marles R, Farnsworth N. Anti-diabetic plants and their active constituents. *Phytomedicine.* 1995; 2:137-189.
103. Singh R, Niaz M, Ghosh S. Hypolipidemic and antioxidant effects of commiphora mukul as an adjunct to dietary therapy in patients with hypercholesterolemia. *Cardiovasc Drugs Ther.* 1994; 8:659-664.
104. Iwu M. Handbook of African medicinal plants. Boca Raton: CRC Press; 1993.
105. Lindhorst K. Antioxidant activity of phenolic fraction of plant products ingested by the Maasai. [MSc. thesis]. Montreal, Canada,

McGill University; 1998.

106. Johns T, Mahunnah R, Sanaya P, Chapman L, Ticktin T. Saponins and phenolic content in plant dietary additives of a traditional subsistence community, the Batemi of Ngorongoro District, Tanzania. *J Ethnopharmacol.* 1999; 66:1-10.

107. Maletnlema T. A Tanzanian perspective on the nutrition transition and its implication for health. *Public Health Nutr.* 2002; 5:163-168.

108. Gibson RS. Traditional methods for food processing diet modification and diversity to increase micronutrient availability. *Ann Nutr Metab.* 2005; 49:10.

109. Kant A, Schatzkin A, Ziegler RG. Dietary diversity and subsequent cause-specific mortality in the NHANES I epidemiologic follow-up study. *J Am Coll Nutr.* 1995; 14:233-238.

110. Hatloy A, Torheim L, Oshaug A. Food variety - a good indicator of nutritional adequacy of the diet? A case study from an urban area in Mali, West Africa. *Eur J Clin Nutr.* 1998; 52:891-898.

111. Johns T. The Chemical Ecology of Human Ingestive Behaviors. *Annu Rev Anthropol.* 1999; 28:27-50.

112. Onyango A, Koski K, Tucker K. Food diversity versus breastfeeding choice in determining anthropometric status in rural Kenyan toddlers. *Int J Epidemiol.* 1998; 27:484-489.

113. Johns T, Sthapit B. Biocultural diversity in the sustainability of developing-country food systems. *Food Nutr Bull.* 2004; 25:143-155.

114. World Health Organisation. Diet, nutrition and the prevention of chronic diseases. Report of a Joint WHO/FAO Expert Consultation. Geneva; 2003: 1-148.

3. INVESTIGATION OF DIETARY INTAKE AND HEALTH STATUS IN EAST AFRICA IN THE 1960s: A SYSTEMATIC REVIEW OF THE HISTORIC *OLTERSDORF COLLECTION*

3.1. Background and objectives for the systematic review of the Oltersdorf Collection

The challenges facing the people of sub-Saharan Africa today have perhaps never appeared more overwhelming. In addition to HIV/AIDS, poverty, and malnutrition, epidemics of non-communicable diseases (NCDs) are the latest threat to this vast continent. Within the next 20 years, sub-Saharan Africa can expect a three-fold increase in deaths due to cardiovascular disease (CVD) and a near three-fold increase in the incidence of type 2 diabetes.[1] According to the World Health Organization (WHO), NCDs currently account for nearly 80% of deaths in developing countries.[2] This statistic is notable in light of current NCD epidemics across the *developed* world.[3]

Our recent review of the *Oltersdorf Collection*, a compilation of investigations from the 1930s-1960s, collected by the *Max-Planck Nutrition Research Unit* in Bumbuli, Tanzania, has revealed that cultural groups throughout East Africa (including Kenya, Tanzania, and Uganda) indeed have a rich history of diversified foods and food habits.[4] Many of the traditional foods identified, including sorghum spp.,[5] various millets,[6,7] bambara groundnuts,[8] root and tubers,[9] green leafy vegetables,[10-13] plant oils[14-17] and wild meats[18] have known health benefits, according to the latest empirical evidence.[4] Moreover, food preparatory practices during this period, including boiling and occasional roasting have been proven to be superior with regard to health-related benefits as compared to current practices in East Africa which favor frying or deep-frying.[19, 20] Overall, the data extraction[4] has provided reason to believe that a healthy, diversified diet may have been possible for the people of East Africa during this time period (1930s to 1960s).[4]

Although Chapter 2 presents and discusses the many health benefits of the traditional East African diet,[4] sub-Saharan Africa continues to experience a *nutrition transition* whereby traditional, well-tried foods and food habits have been systematically replaced with the manufactured food products of the multinational corporations. This *globalized food culture* has been implicated in the genesis of NCDs, including type 2 diabetes, obesity, and CVD, throughout the world.[2124]

Recent evidence has suggested that the *nutrition transition* in East Africa continues to be fueled by: (1) the exploitation of arable land for the production of cash crops for the western market economies,[25, 26] (2) the environmental degradation of arable land for economic advancement, (3) the economic dependence (i.e. national debt) manufactured and incurred through such organizations as the International Monetary Fund (IMF) and World Bank,[27, 28] (4) the forced, rapid urbanization of rural populations for reasons of economic survival,[29-31] (5) the introduction of patented, genetically-engineered crops by the controversial biotechnology company, Monsanto[32, 33] and (6) the monopolization of the *globalized* food system by a few multinational corporations, including: *Nestle©, Altria©, Unilever©* and *Cocacola* ©.[34, 35]

Recent empirical investigations from Okinawa, Japan,[36, 37] the Mediterranean[38, 39] and China[40, 41] have provided robust evidence that the adherence to traditional, culture-specific dietary patterns is positively associated with health status, including reduced prevalence of nutrition-related diseases, and increased lifespan. This empirical evidence has been supported by the largely suppressed research of Dr. Weston Price.[42, 43] Price conducted his research among fourteen distinct native populations including the inhabitants of the Outer Hebrides of Scotland, the Canadian Arctic, New Zealand, Australia, the Polynesian Islands, and Africa and was astonished by the excellent physiques and health status indicators of these groups, which included, in African tribes, a resistance to infectious diseases.[43] Traditional dietary patterns, he concluded, were a key determinant of the remarkable levels of health status.[42, 43] Clearly, given the global epidemics of obesity and type 2 diabetes today, investigations of traditional food habits must continue as these investigations are likely to hold the key to abating these epidemics.

The *Oltersdorf Collection* represents a compilation of the earliest available nutrition-related research conducted throughout East Africa (i.e. Tanzania, Kenya, and Uganda) from the 1930s to the 1960s.[4] The purpose of the present investigation was to determine if relationships between dietary patterns and indicators of health status were investigated in specific cohorts in East Africa during this time period (1930s to 1960s). The overall objectives were three-fold:

(1) To determine if the dietary intake was adequate and consisted of a diversity of traditional whole foods representative of the wide spectrum of food choices available in the region at this time;[4]

(2) To evaluate and report on the prevalence of NCDs, including obesity, type 2 diabetes, and hypertension, and other health-related indices (e.g. anthropometric, biochemical, and clinical assessments including prevalence of parasitic infections) in the specific cohorts studied; and

(3) To determine if relationships between dietary intake/adequacy and health status were investigated and quantified to provide empirical evidence of the health-related benefits of East African food habits at this

time.

3.2. Methods of the systematic review
3.2.1. Oltersdorf Collection

The *Oltersdorf Collection* consists of 75 reports collected by the *Max-Planck Nutrition Research Unit*, previously located in Bumbuli, Tanzania (formerly Tanganyika).[4] Written in both English and German, these reports include observational research investigation manuscripts, research records, and books produced from nutrition-related studies conducted throughout Kenya, Uganda, and Tanzania, including Zanzibar Island and Pemba Island, from the 1930s to 1960s. All reports included within the *Oltersdorf Collection* have been manually scanned and converted to full text PDF-documents which are now accessible for download *via* the online collection (which was available at: http://www.healthyeatingclub.com/Africa/ and has been transferred to http://www.dr.verena.com)[44] The purpose of this online collection is to present data related to the foods and food habits of the peoples of East Africa during this time period,[44] and these data have been summarized recently in the *Asia Pacific Journal of Clinical Nutrition*.[4]

3.2.2. Data extraction and classification

The 75 reports comprising the *Oltersdorf Collection* were manually reviewed by the principal researcher (V.R.). Reports were included in the present review if they included all of the following:

(1) Name and geographic region of the specific cohort investigated;

(2) Demographic descriptors of the cohort investigated, including the gender, age, and number of individuals studied; and

(3) Presentation of original, quantitative data related to both dietary

intake/adequacy *and* health status. Dietary intake and adequacy may have been quantified, for example, using dietary recall assessments, weighted food records, and/or dietary questionnaires (e.g. food frequency questionnaires). Health status may have been quantified, for example, using data pertaining to anthropometric, biochemical, and clinical assessments including the assessment of non-communicable and infectious diseases prevalence.

Reports that did not fulfill these criteria were excluded.

3.3. Results of the systematic review
3.3.1. Systematic review process

A flow diagram of the systematic review process is presented in Figure 2. Of the 75 reports comprising the *Oltersdorf Collection*, 56 were excluded due to a lack of original, quantitative data related to dietary intake or health status. These included 22 review papers,[45-66] 20 qualitative reports on foods and food habits,[67-86] six reports on the chemical analysis of foods,[87-92] three reports on agricultural practices,[93-95] two reports on food markets,[96, 97] two reports on nutritional policy,[98, 99] and one report presenting food balance data.[100]

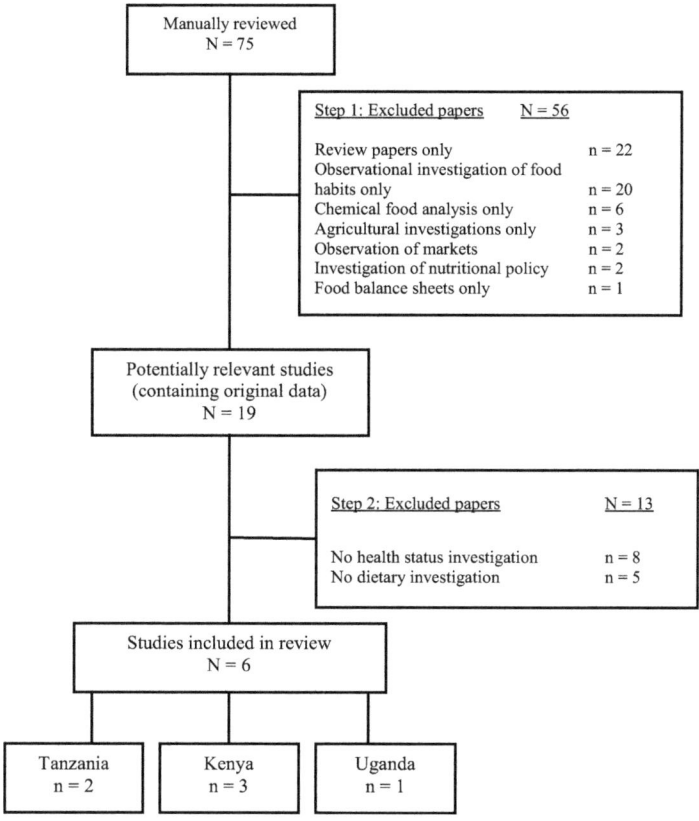

Figure 2. Flow of reports of the *Oltersdorf Collection* included/excluded from review.

Nineteen reports contained original data. Of these 19 reports, 13 were excluded,[46, 101-112] including eight reports which evaluated dietary intake/adequacy but did not evaluate health status, and five reports which evaluated health status but did not evaluate dietary intake/adequacy (Figure 2).

The systematic review resulted in six original investigations presenting data related to dietary intake *and* health status within a specific cohort residing in East Africa (Figure 2).[112-117] These reports present investigations of various ethnic groups of Tanzania,[113, 114] Kenya[112, 115, 116] and Uganda.[117] Few if any empirical investigations have evaluated dietary patterns and health status indices in East Africa before this period.[118]

3.3.2. Limitations of the research

While these reports may be valuable as the earliest nutrition-related investigations conducted in East Africa, none of these investigations performed statistical analyses (e.g. correlations and regressions) between measures of dietary intake and measures of health status. Further, in studies investigating more than one ethnic group or performing repeated measures over time, the differences *between* the specific groups and *within* the specific group over time were never evaluated statistically.

3.3.3. Funding sources

Studies conducted by Popleau *et al.*[114] and Kreysler *et al.*[113] in Tanzania were funded by the *Max-Planck Nutrition Research Unit*, which was operated under an agreement with the government of Tanzania and was supported financially by the following German Foundations: "Brot für die Welt" (Bread for the world), the *Fritz Thyssen Foundation* and the *Robert Bosch Foundation.* Studies conducted by Keller *et al.*[115] and Korte *et al.*[112] in Kenya were funded by the *Institute for Human Nutrition* and the *Tropical Institute of the Justus-Liebig-University* in Gieβen, Germany, in cooperation with the Kenyan Government. Studies conducted by Bohdal *et al.*[116] throughout Kenya were funded by the *World Health Organization*. Studies by Jelliffe *et al.*[117] conducted in Uganda were funded by the

Department of Paediatrics and Child Health and the *Department of Preventive Medicine* at the Makerere Medical School in Kampala and by the *Virus Research Institute,* located in Entebbe, Uganda.

3.3.4. Overview of the research

Survey areas and ethnic groups

In Tanzania, all investigations were conducted in the North-East.[113, 114] Popleau *et al.*[114] investigated groups residing in the Usambara Region, NorthEastern Region and Coastal Region, while Kreysler *et al.*[113] investigated groups residing in the high- and lowlands of Kilimanjaro, NorthPare, South -Pare and the Usambara Regions (Table 7). In Kenya, investigations were conducted in the Nyeri[115] and Mwea-Tebere villages[112] and in the Central, Nyanza and Eastern Provinces (Table 8).[116] In Uganda, investigations were conducted in both the rural and urban regions of the Acholi district (Table 9).[117] The distinct ethnic groups studied in the various regions are presented in parentheses in Tables 7-9.

Time period and duration

All investigations were conducted between 1960 and 1967.[112-117] The data collection process was generally completed within 4 months, with longer periods being reported by Kreysler *et al.*[113] (6 months), Bohdal *et al.*[116] (6 months), and Popleau *et al.*[114] (12 months).

Sample sizes

Sample sizes were provided for all investigations except those conducted in the village of Soni, Tanzania[114] and in the Ichuga and Kiamariga villages of Kenya[115] where the number of families were stated in place of subject numbers. Sample sizes ranged from 34 to 1190 subjects, and are presented in Tables 7-9.

Age

All investigations provided an age range. Age ranged from 0-70 years with the exception of one survey conducted by Jelliffe *et al*,[117] where the age range was 0-3 years.

Gender

Gender distributions were reported in all investigations except one, which was conducted in Uganda with children only.[117] The number of female subjects was only slightly higher than the number of male subjects in all the investigations conducted in Tanzania[113, 114] and Kenya.[112, 115]

Seasonal variation surveys

An investigation of seasonal variation changes conducted in the Mwea-Tebere villages in Kenya[112] provided an example of the impact that seasonal variations can have on the dietary patterns and dietary intake within a specific cohort.

Assessment of dietary intake and adequacy

In Tanzania, dietary intake and adequacy were evaluated by means of a 24-hour recall combined with a 4-day weighted food record,[114] and a 24-hour recall combined with a food frequency questionnaire (FFQ).[113] In Kenya, investigations evaluated dietary intake and adequacy by using a FFQ in combination with a 7-day weighted food record.[112, 115, 116] In Uganda, dietary intake and adequacy were evaluated by using a dietary questionnaire and a 7-day weighted food record.[117] Most investigations evaluated dietary intake within the family unit.[112-117] Jelliffe et al.[117] investigated dietary intake within infants and children only. Outcomes of the assessments related to dietary intake and dietary adequacy are presented in Tables 7-9.

Assessment of health status

The majority of investigations (5/6, 83%) evaluated anthropometric, clinical and biochemical measures as indicators of health status.[112-114, 116, 117] Keller et al.[115] evaluated anthropometric and clinical assessments only. Outcomes of the health status assessments are presented in Tables 7-9.

3.3.5 Outcomes related to dietary intake and adequacy

Staple foods & beverages

In Tanzania, white maize was reported as a primary staple food in all regions except the coastal areas of Tanzania, where cassava and fish served as the main staples (Table 7).[113, 114] Maize was also reported as a primary staple food in some areas of Kenya, including the Nyeri villages,[115] the Mwea-Tebere villages,[112] the Ngamwa villages, and in the second survey conducted in the Masumbi village (Table 8).[116]

The survey in Uthiuni village in the Eastern Province of Kenya was conducted during a period of maize shortage when the people were receiving imported yellow maize as famine relief. Due to the maize shortage there was a higher consumption of wheat flour in the Uthiuni village compared to other areas (Table 8).[116] Yellow maize was also distributed as a famine-relief food during the second survey carried out in the Nyaani village (Eastern Province, Kenya), where millet and sorghum ordinarily served as the primary staple foods (Table 8).[116]

Sorghum was a primary staple food in the first survey conducted in Masumbi village among the *Luo* (Eastern Province of Kenya) (Table 8).[116] In Uganda, eleusine millet (*kwon*) and maize were the primary staple foods among infants in rural Acholiland and urban Kampala, respectively (Table 9).[117]

Sweetened tea with small amounts of milk, stiff maize porridge (*ugali*), and thin maize soup (*uji*) for breakfast were documented as staple foods/beverages in all investigations in Tanzania and Kenya (Tables 7 and 8).[112-116] The quantity and varieties of food consumed during afternoon and evening meals varied somewhat between the areas surveyed and changed slightly with the seasons. *Ugali* made from cereals (mainly maize) served with a legume and green vegetable mixture represented the most commonly consumed meal in most surveys conducted in Tanzania and Kenya.[112-116]

In rural Acholiland, Uganda, prolonged breast feeding was performed in combination with an early introduction of weaning foods based on vegetable proteins, including an eleusine millet gruel *(kwon)*, sesame, and varieties of legumes (i.e. *Phaseolus vulgaris, Vigna Unguiculata, Cajanus indicus)*.[117] Urban children in Kampala were fed maize flour and other carbohydrate foods (Table 9).[117]

Dietary protein

In Tanzania, protein intake was highest in the coastal areas (Chakichani 54% of dietary intake, and Maranzara 27% of dietary intake), probably due to regular fish consumption (Table 7). However, children in the coastal regions were fed cassava rather than fish.[114] A high intake of milk (15% of the total protein intake) was noted in Leguruki compared to villages of the Usambara region where animal protein intake (13-15%) was derived from meat, dried and fresh fish, and organs (Table 7).[114] According to Kreysler et al.[113] the lowlands of Tanzania had a higher protein intake, in particular milk and fish intake, as compared to the highlands (Table 7).

In Kenya, meat intake was reportedly consumed only once or twice per month, while milk was regularly used in small amounts in tea and *uji* (thin soup made out of maize flour) among the Kikuyu farmers in the Nyeri district villages (Table 8).[115] In the Mwea-Tebere villages, protein consumption was below recommended values.[112] Protein intake was also determined to be low in the Ngamwa village of the Central Province, in the first and second survey in the West Koguta village of the Nyanza Province, and in Uthiuni village of the Eastern Province. In these regions, protein was primarily derived from cereals followed by legumes (i.e. Ngamwa and Uthiuni villages) or fih (i.e. West Koguta) (Table 8).[116] Increased consumption of animal protein was observed during the second survey of the West Koguta village as compared to the first survey (38% vs. 25%, respectively, due to a higher consumption of fish), and during the second survey of the Nyaani village in the Eastern Province of Kenya (probably due to the distribution of tinned pork by famine relief) (Table 8).[116] In the pre-harvest seasons of the Central, Nyanza and Eastern Provinces of Kenya, food intake was precarious in many communities and the protein score of most families fell below 50% adequacy.[116] The proportion of protein from animal sources was highest and well above recommended values due to the regular fish intake in both surveys conducted in the Masumbi village of the Nyanza Province (Table 8).[116]

Dietary fat

In Tanzania, fat represented a small proportion of the diet in all survey areas (Table 7).[113, 114] In Kenya, increased consumption of cheap vegetable oils was observed at the end of the dry season (March) in the Mwea-Tebere villages (Table 8). The oil was used to improve the taste of disliked foods. Of the surveys conducted by Bohdal et al.,[116] dietary fat was derived from ghee and unspecified vegetable oils.

Caloric intake

Caloric intake was adequate in 95% of the villages surveyed by Popleau et al.[114] in North-East Tanzania, including the Usambara region, NorthEastern region and the coastal region (Table 7). This was also the case in Kenya, in the Nyeri District and the Mwea-Tebere villages (Table 8). In Ngamwa village of the Central Province of Kenya the food stores were exhausted by June (at the time of the survey) and 54.2% of the families consumed less than 80% of the recommended caloric intake.[116] Investigation of the diet in the West Koguta village of the Nyanza Province near Lake Victoria showed caloric adequacy after the yearly harvest (first survey), and inadequacy during the second survey conducted six months later (Table 8). Masumbi village in the Nyanza Province of Kenya had two harvests per year. Both surveys in Masumbi were conducted after these harvests and showed a moderate number of cases below caloric adequacy (29.2% in the first survey and 18.2% in the second survey).[116] In Uthiuni village in the Eastern Province of Kenya, results were obtained long after the main harvest and 59.1% of the families were below recommended caloric intake.[116] Nyaani village of the Eastern Province of Kenya was hit by drought and was under famine relief during the second survey, during which yellow maize and tinned pork were distributed and consumed.[116] This resulted in a decrease in families being below 80% caloric adequacy from pre (23.8%) to post (14.3%) famine relief.[116]

3.3.6. Outcomes related to health status

Prevalence of non-communicable diseases

No reports of obesity, hypertension, or type 2 diabetes were presented in any of the investigations conducted in East Africa during this period.[112-117]

Anthropometric data

Developmental deficiencies in infants and children, including muscle wasting and low weight and height, were observed in studies conducted in North-East Tanzania.[113, 114] Such deficiencies were also observed in the Eastern Province of Kenya where signs of kwashiorkor or marasmic-type of muscle wasting were identified in several children.[116] A deviation toward lower anthropometric values was observed in children within the Mwea-Tebere villages of Kenya.[112] An investigation by Keller *et al.*[115] conducted in the Nyeri district villages of Kenya revealed little to no developmental deficiencies in children. In Uganda, urban newborns and children (1-3yr olds) were more likely to have a normal weight and less malnutrition as compared to their rural counterparts.[117]

Protein-energy malnutrition

In Tanzania, dyschromotrichia of the hair (11.5%) followed by parotid gland enlargement (10.6%) were the most frequent diagnoses related to protein-energy malnutrition as revealed by Kreysler *et al.*[113] (Table 7). By contrast Popleau *et al.*[114] reported no severe signs of protein-energy malnutrition and no cases of kwashiorkor in North-East Tanzania.

In Kenya, depigmentation of the hair was the most common protein deficiency observed in children (Table 8).[112, 115] The surveys conducted by

Bohdal et al.[116] in the Nyanza and Eastern Provinces of Kenya revealed a high prevalence of protein-energy malnutrition as reflected by such clinical observations as parotid enlargement, thinness of hair, hepatomegaly, dyspigmentation of skin, lack of lustre, and pallor of membranes (Table 8). In the Central Province, low incidence of protein-energy malnutrition was observed.[116]

In Uganda, diseases related to protein-energy malnutrition, such as kwashiorkor (0.4%) and hypochromotrichia (4.6%) were not highly prevalent among children living in rural Acholiland. By contrast, urban Acholi children showed a higher prevalence of kwashiorkor (9%) and hypochromotrichia (22.8%) (Table 9).[117]

Popleau et al.[114] determined that plasma protein concentrations were within the normal range (7.27g/100ml ± 0.61) in all areas which they surveyed in North-East Tanzania (Table 7). In Kenya, assessment of the total protein in serum in the Mwea-Tebere villages revealed that no severe deviation from normal values existed on average.[112]

Parasitic infections

Common symptoms and/or diseases related to parasitic infections in North-East Tanzania included splenomegaly and hepatomegaly with a higher prevalence in the lowlands (19.4%) as compared to the highlands (0.3%).[113] Malaria, attributed to the protistan parasite of genus *Plasmodium,* was reported in the Nyanza (West Koguta and Masumbi villages) and Eastern Province (Uthiuni and Nyaami villages) of Kenya.[116] These cases were associated with splenomegaly, hepatomegaly and anemia (Table 8).[116] Splenomegaly and hepatomegaly were also caused by infections of *Ascaris lumbricoides* in the Masumbi village (Nyanza Province) and *Visceral leishmaniasis* (i.e. *black fever*) in the Nyaami village (Eastern Province) of Kenya.[116] In Uganda, malaria was observed among the rural Acholi children, as reflected by the number of enlarged livers and spleens observed (Table 9).[117]

Helminth infections were evaluated in most regions[112-114, 116] with *Ascaris lumbricoides* being the most frequently reported parasitic infection (2-80% prevalent) (Table 7 and 8). In North-East Tanzania, the incidence of *Ascaris lumbricoides* infections was lowest in the two coastal villages, Maranzara (2%) and Chakichani (9%), and highest in Leguruki and Mulungui villages (both 80%) (Table 7).[114] The incidence of *Ascaris lumbricoides* and hookworm (*Necator americanus* and *Ancylostoma duodenale*) infections seemed to be higher in the highlands (51%; 16%) of North East Tanzania as compared to the lowlands (14%; 5%).[113] Korte et al.[112] revealed that *Ascaris lumbricoides* infection was common among children 1-4 years of age, and bilharziasis caused by *Schistosoma mansoni* was a major concern among school-aged children.[112]

Anemia

In North East Tanzania, haemoglobin, haematocrit values and mean cell volume haemoglobin content (MCHC) were within the normal ranges.[114] On average, higher haematological values were observed in the lowlands as compared to the highlands.[113]

Haematological data collected by Bohdal et al.[116] revealed anemia related to parasitic infections of African trypanosomiasis in the Nyanza and Eastern Provinces of Kenya, and *Visceral leishmaniasis* in the Nyaami village of the Nyanza Province of Kenya.[116]

Jelliffe et al.[117] evaluated a high prevalence of anemia among the rural Acholi children (Table 9).

Vitamin C adequacy

Haematological vitamin C (ascorbic acid) concentrations were determined to be adequate in Tanzania, as reported by Popleau et al.[114] and Kreysler et al.[113] (Table 7).

Vitamin A adequacy

Dietary intake of vitamin A was observed to be deficient in several investigations, as reflected by the prevalence of the following clinical conditions: Follicular hyperkeratosis (Range: 7-12%),[112-115] xerosis conjunctivae (Range: 3.2-29.1%),[113, 116] xerosis of skin (30%)[112] and corneal vascularisation (Range: 9.5-19.3),[112, 116] (Table 7-9).

Investigations conducted by Kreysler et al.[113] in North-East Tanzania revealed that the prevalence of clinical vitamin A deficiency was higher in those residing in the lowland versus the highland areas. Biochemical vitamin A deficiency, as reflected by low serum retinol and carotene concentrations was observed in North-East Tanzania.[113] However, these indices were determined to be within the normal range in the Mwea-Tebere villages of Kenya.[112]

Riboflavin and niacin adequacy

Intake of niacin and riboflavin was found to be inadequate in the studies conducted in North-East Tanzania by Popleau et al.[114] Biochemical analyses revealed riboflavin insufficiency in all regions of North-East Tanzania except the Kilimanjaro highlands and the South Pare region, where a remarkably good riboflavin status was evaluated.[113] The most frequently reported disease related to inadequate intake of these micronutrients in Tanzania was atrophic lingual papillae (8.3% prevalent in the study by Kreysler et al).[113]

In Kenya, dietary intake of riboflavin in all survey areas was below recommended levels as reflected in the high prevalence of atrophic lingual papillae and angular palpebritis but stomatitis angularis was the main prevalent sympotom.[112, 115, 116]

3.4. Discussion

The purpose of the present investigation was to perform the first systematic review of the historic *Oltersdorf Collection*[4] to determine if relationships between traditional dietary food habits and indicators of health status were investigated within specific cohorts residing in East Africa from the 1930s to 1960s. The overall objectives were: (1) to determine if the dietary intake was adequate and consisted of a diversity of traditional foods, representative of the wide spectrum of traditional whole foods available throughout the region at this time;[4] (2) to evaluate and report on the prevalence of NCDs and other health-related indices in the specific cohorts studied; and (3) to determine if relationships between dietary intake/adequacy and health status were investigated and quantified to provide empirical evidence of the health-related benefits of East African food habits at this time.

The review process resulted in six investigations being identified.[112-117] Published between 1963 and 1969, these are probably the first investigations to provide original data regarding to dietary intake/adequacy and health status indices within specific East African cohorts (Table 7-9).

3.4.1. Dietary intake and adequacy

Overall, the present review has demonstrated that dietary deficiencies existed amongst several of the cohorts investigated, and that the dietary intake did *not* consist of a wide spectrum of traditional foods available in the region at this time.[4] The investigations overwhelmingly documented a monotonous diet *lacking* in dietary diversity.[112-117] The main overall finding of the dietary intake surveys was the high cereal consumption, primarily white maize, which routinely provided over 50% of the total caloric intake.[112-117]

There are many potential explanations for the lack of dietary diversity in the cohorts studied. Several investigators have discussed the transition from the traditional East African diet consisting of roots, tubers, legumes, wild game meat, fish, milk, and cereals including sorghum and varieties of millets, to a so-called *maize* diet.[116, 118-120] The shift toward maize as a primary staple food in the majority of East African diets began with the emergence of cash-crop farming during the 1950s.[121-123]

Indigenous communities were encouraged by colonial masters to grow cash crops for export to "improve their standards of living" within a new economic system.[28, 121] The transition to cash-crop farming had significant consequences on the nutritional status of the East African population.[123] For example, the clearing of native forest for cash-crop farms eliminated many indigenous food trees, wild fruits and vegetables, and gathered foods from the diet.[124-127] Moreover, the ecological damage made hunting virtually impossible and markedly reduced the amount of wild bush meats consumed by the local populations.[121]

The expansion of cash-crop farming of coffee, copra, cotton, sesame, peanuts, and sugar in East Africa, also reduced the number of cattle farms, which in turn displaced domesticated animal and milk protein in the diet.[25, 28, 116] Inevitably, this resulted in a monotonous cereal-based diet, devoid of animal foods[128] as reflected in the monotonous diets reviewed (Table 7-9). The transition to a cash-crop economy also had a negative effect on the cultivation of the less commercially acceptable indigenous crops, such as protein-rich sorghums and millets, that had previously contributed to the dietary diversity of the diet of the indigenous people.[32, 122, 126]

Since colonization, the loss of arable land, the destruction of natural ecosystems, and the replacement of indigenous crops with cash crops has resulted in extended periods of food shortage and an increased dependence on introduced or donated cereals.[122, 129, 130] For example, research by Bohdal et al.[116] in the Uthiuni village of Kenya (Table 8) was conducted while the local population was experiencing a famine (during survey 8). Foods such as tinned pork and maize were normally not included into the diet, but were consumed only as food aid during the period of food shortage. In this example, maize replaced the nutritionally superior millets as the primary staple grain. It is noteworthy that the food aid provided to the Uthiuni village during the famine was insufficient to markedly improve the health status of this population.[116]

The decline in the use of indigenous food crops and gathered wild species, and the reliance on food-aid programs, delivering wheat, rice and maize resulted in the overall simplification of the African diet.[21, 131-133] Interestingly, recent research and epidemiological evidence has associated the process of *westernization* of dietary habits and traditional food habits with the increasing prevalence of NCDs among indigenous populations.[134-138] As has already be mentioned, the rise of NCDs adds to the existing health burden of micronutrient deficiencies and infectious diseases throughout Africa.

Indeed the loss of dietary diversity is not limited to East Africa. Dietary simplification has become a global issue, as greater numbers of people are being fed with fewer and fewer crop species.[139] It is generally accepted that as societies develop, they experience a reduction in the number and variety of food sources consumed.[22, 140] Therefore, with the passage of time the knowledge of traditional foods and food habits is lost.[130] Dietary simplification and the loss of ancient knowledge regarding traditional foods and food habits continues today, largely unabated in East Africa.[19, 124] Oniang'o[141] and Kuhnlein[130] assert that indigenous food resources and the knowledge of their uses has greatly diminished in sub-Saharan Africa. However, uncultivated food resources are still gathered and used, particularly by rural communities, when other foods fall short.[142, 143]

3.4.2. Health Status

Non-communicable diseases (NCDs)

The present systematic review provided no mention of NCDs in any cohort investigated in East Africa during the 1960s. Obesity, type 2 diabetes, CVD and hypertension were not mentioned as clinical diagnoses in any study reviewed.[112-117] It must also be mentioned however, that all of these investigations were conducted several years before the widespread dissemination of manufactured food products of the transnational corporations, which has hastened NCD epidemics markedly. Several researchers have attributed the current rise of NCDs in Africa to the introduction of these manufactured food products and poor staple foods including cheap vegetable oils and wheat.[137, 138, 144, 145]

Malnutrition-related diseases

Overall, children were the most affected by the change from a diverse traditional diet toward a more simplified diet, as reflected in the high incidence of clinical, anthropometric and biochemical deficiencies indicating protein-energy malnutrition.[112-117] The replacement of protein-rich foods, such as traditional millet porridge with sweetened tea for breakfast (Table 7-9) was particularly detrimental to children's health.[120] Servere protein-calorie malnutrition, including kwashiorkor, was uncommon among the young children in rural Acholiland, reflecting the importance of prolonged breast feeding and the availability of traditional weaning foods for the physical development of infants.[117]

The widespread prevalence of micronutrient malnutrition, such as vitamin A deficiency, iodine deficiency, and anemia, was observed in several investigations reviewed (Tables 7-9).[112-117] These deficiencies remain in the forefront of major nutritional diseases in East Africa today.[146, 147] As adequate intake of vitamin A protects against infections of the respiratory tract, and reduces the risk of transmission of HIV infection from mother to child, the alleviation of vitamin A deficiency is particularly important.[148-151]

Kreysler et al.[113] suggested that replacement of dark green leafy vegetables by European cabbage deprived the people of important carotene sources, resulting in the high prevalence of vitamin A and riboflavin deficiencies.[113, 114] In North-East Tanzania, vitamin C adequacy was attained with regular consumption of fruits, wild vegetables and tubers. However, based on the dietary questionnaire conducted by Kreysler et al.,[113] a tendency to replace these staples with maize and rice was observed, particularly among the inhabitants of the lowlands.[113] Loss of several varieties of crops which the land historically yielded (e.g. sorghum and millet varieties, bambara groundnut, beans), and which were high in micronutrients, resulted in reduced biodiversity and hastened dietary simplification.[129, 152]

Unfortunately, programs of the United Nations (UN) and the aid agencies which are mostly directed at children use sugar, harmful oils and fats, and monosodium glutamate for vitamin A fortification.[148, 153] The high intake of these particular food constituents is known to hasten the development of overweight/obesity and related NCDs, including type 2 diabetes and CVD.[22, 154] While fortification (and vitamin supplementation) are common strategies for alleviating micronutrient deficiencies, as they are cost-effective and easy to administer, they are inadequate as they do not address the underlying cause of the deficiency, including inequity and poverty.[155] Little emphasis has been placed on whole food approaches to control micronutrient malnutrition.[156] Existing data has provided an insight into how whole foods can significantly contribute to better health status.[4, 17, 157-161]

Infectious diseases

Several studies noted a high incidence of parasitic infections amongst the population groups investigated in North-East Tanzania, Kenya, and Uganda (Table 7-9). Protein-energy malnutrition and anemia co-existed with a high incidence of helminth infestations, and malaria parasitemia among the Kenyan population groups investigated by Bohdal *et al.*[116] Jelliffe *et al.*[117] reported that the high anemia rates among the rural Acholi children were due to the high malaria incidence, an associated link also reported by Nyakeriga *et al.*[162] Population density, low standards of hygiene and malnutrition were the fundamental determinants of the high incidence of *Ascaris lumbricoides* observed among all population groups of North East Tanzania[113, 114] and Kenya.[112, 115, 116]

3.4.3. Relationships between dietary intake/adequacy and health status

Relationships between dietary intake/adequacy and health status indices were not statistically investigated in any of the studies reviewed.[112-

[117] It is not possible, therefore, to provide robust empirical evidence of the healthrelated benefits of East African food habits at t his time.

In summary, the systematic review of the *Oltersdorf Collection* revealed that many ethnic groups investigated in East Africa (i.e. Tanzania, Kenya and Uganda) during the 1960s did not exhibit adequate dietary intake and did not consume a diversity of traditional whole foods representative of the wide spectrum of food choices available in the region at this time.[4] While NCDs were not prevalent, there was substantial reporting of malnutrition-related and infectious diseases, particularly prevalent among children. The present review supports the contention that a shift from the traditional, diversified diet[4] to a more simplified monotonous diet may have been concomitant with the implementation of cash-crop farming. For a resolution of current nutrition-related epidemics currently plaguing the vast continent of Africa (i.e. NCD and malnutrition-related diseases; the *double burden*) it is critically important to investigate and disseminate evidence regarding the fundamental contributors to the *nutrition transition*.[21]

3.5. Résumé

We have recently reported on the myriad of health benefits of traditional East African food habits. However, this region continues to experience a *nutrition transition* whereby traditional, well-tried foods and food habits have been systematically replaced with the products of multinational corporations. The health-related impact has been devastating, as clearly evidenced by current non-communicable disease (NCD) epidemics. The purpose of the present investigation was to review the *Oltersdorf Collection* and determine if relationships between dietary intake/adequacy and indicators of health status were investigated in specific cohorts in East Africa (i.e. Tanzania, Kenya and Uganda) from the 1930s and 1960s. The systematic review process resulted in six investigations being identified. Published between 1963 and 1969, these are likely the first investigations to provide original data pertaining to dietary intake/adequacy and health status indices within specific East African cohorts. Overall, the review revealed that many ethnic groups did not exhibit adequate dietary intake and did not consume a diversity of traditional whole foods representative of the wide spectrum of food choices available in the region at this time. While NCDs were not prevalent, there was substantial reporting of malnutrition-related and infectious diseases, particularly among children. The present review supports the contention that the shift from a traditional, diversified diet to a simplified, monotonous diet may have been concomitant with the onset of cash-crop farming. For resolution of the current nutrition-related epidemics currently plaguing the vast continent of Africa (i.e. NCD and malnutrition-related diseases; the *double burden*), it is critically important to investigate and disseminate evidence related to the fundamental contributors to the *nutrition transition*.

Table 7. Investigations in North-East Tanzania.

Authors, year Survey area (*Ethnic group*) Sample (n)	Dietary intake	Dietary adequacy	Health status measures and prevalence of diseases

Popleau et al.[114] 1969
North-East Tanzania:
Usambara region:
○ Bumbuli village(*Shambala*)
n = 238
○ Mulungui village
n = 85
○ Soni village (*Shambala*)
n = 20 families
North-Eastern region:
○ Leguruki village (Meru)
n=34
Coastal region:
○ Maranzara village (*Digo*)
n=35
○ Chakichani village (*Digo*)
n=35
	Usambara region:
	• Most important staple food: white maize and plantain
	• Supplementary foods: wild spinach, cassava flour, legumes, meat, fish organs, milk
	• Total intake of animal protein: 13-15% from meat, fish, organs
	
	North-Eastern region:
	• Most important staple food: white maize
	• Supplementary foods: legumes, milk, meat, blood
	• Total intake of animal protein: 24%
	  - 15/24% came from milk
	  - 9/24% came from meat and blood
	
	Coastal region:
	• Most important staple food: fish, cassava
	• Supplementary foods: maize, wild spinach
	• Total intake of animal protein:
	  Chakichani 54%
	  Maranzara 27%
	
	• In all areas with exception of Soni and Chakichani legumes were an important protein source
		Notes from all areas investigated:
		• Caloric requirements were met by 95% of the individuals studied
		• Calcium requirement were met by less than 60%
		• Diets were overall inadequate with respect to intake of protein, vitamin B_2, niacin and vitamin A
			Anthropometric:
			Infants/Children < 2 yr of age
			• Muscle wasting prevalent
			• Uniformly delayed increase in height and weight
			Note: Physical development of infant was considerably better in Leguruki compared to villages of the Usambara area
			
			Clinical:
			• No severe signs of protein-energy malnutrition
			• No cases of kwashiorkor
			• Follicular hyperkeratosis: 9.8% (evenly distributed over all areas examined)
			
			Parasitic (%) infections:
			
			          A. lumbricoides Hookworm
			Bumbuli 63% 11%
			Mulungui 80% 28%
			Soni 40% 21%
			Leguruki 80% 17%
			Maranzara 2% 14%
			Chakichani 9% 38%
			
			Biochemical:
			• No cases of microcytic anemia
			• Total plasma protein (g/100ml): 7.27±0.61
			• Ascorbic acid (mg%): 1.26±0.36

Table 7 (continue). Investigations in North-East Tanzania.

Authors, year Survey area (*Ethnic group*) Sample (n)	Dietary intake	Dietary adequacy	Health status measures and prevalence of diseases

Kreysler et al.[113] 1969
North-East Tanzania:
Kilimanjaro region:
○ Highlands (*Chagga*)
○ Lowlands (several tribes)
 	
North-Pare region:
○ Highlands (*Pare*)
○ Lowlands (*Pare*)
 	
South-Pare region:
○ Highlands (*Pare*)
○ Lowlands (*Pare* and *Maasai*)
 	
Usambara region:
○ Highlands (*Shambala*)
○ Lowlands (several tribes)
 	
Total N=1190
	Breakfast:
	• Most common food: tea with sugar
	• Replaced by *uji** in poor families
	• Lower white maize consumption in lowlands
	
	Lunch:
	• Most important staple food: white maize (increasing importance from Kilimanjaro (Moshi) to Usambara)
	• Highlands: higher banana consumption
	• Lowlands: *Mchicha*** with extensive use of milk (with exception of Moshi)
	
	Supper and lunches
	• Lowlands: high fish consumption
	• Highlands increased consumption of:
	  • Milk (North Pare)
	  • Beans (Kilimanjaro low, South-Pare low)
	
	Prevalent dishes (% of all lunches and suppers):
	• *Ndizi na nyama* (bananas and meat) in Kilimanjaro highlands (37%)
	• *Ugali**** with fish in Usambara Lowlands (40%)
		• Data not reported	Anthropometric (children only):
			• Poor anthropometric status in children up to 10 years
			
			Clinical (% affected):
			• Dyschromotrichia of hair: 11.5%
			• Parotid enlargement: 10.6%
			• Xerosis conjunctivae: 29.1%
			• Follicular hyperkeratosis: 7.3%
			• Atrophic lingual papillae: 8.3%
			• Spleno-/ and hepatomegaly: 10.6%
			
			Parasitic infections:
			• Ascaris lumbricoides: 14-51%
			• Hookworm: 5-16%
			
			Biochemical:
			• Non-physiological anemia prevalent
			• Protein in serum (g%): 4.40-4.42
			• Albumin in serum (g%): 6.72-6.83
			• Retinol in serum (μg/100ml): 28.73
			• Carotene in serum (μg/100ml): 93.72
			• Average riboflavin (μg)/creatinine (g) ratio (adults): 249-300
			• Average ascorbic acid levels (mg/100ml) in blood (adults): 1.02

*Thin soup made out of maize flour; ** Any type of wild green leafy vegetable; ***Stiff maize porridge, mainly made from maize.

Table 8. Investigations in Kenya.

Authors, year Survey area (*Ethnic group*) Sample (n)	Dietary intake	Dietary adequacy	Health status measures and prevalence of diseases

Keller et al. [115]
1969
 	
Nyeri district Villages (*Kikuyu*)
  o Kiamariga village
  o Ichuga village
 	
Total N = 66 families
	Breakfast:
	• *Kiamariga*: Tea with milk, sugar or both and *uji** with salt, sometimes milk or sugar;
	• *Ichuga*: Tea consumed twice a day, as often as *uji**
	
	Lunch/snack:
	• Sweet potatoes, yams, arrow roots with potatoes, roasted maize on cob, a piece of sugar cane or a cup of *uji** or tea
	• *Ugali*** with milk or vegetables
	• Potatoes with vegetables
	
	Supper:
	• Stew of maize, beans and vegetables (single or in combination, including fried onions, cabbage, potatoes, bananas, wild vegetables)
	
	Animal protein:
	• Milk: consumed on regular basis with tea or *uji*
	• Meat: consumed once or twice per month
	• Eggs: rare, delicacy item
	
	Wild vegetables:⁺
	- regular part of diet especially towards the end of rainy season
		Macro and micro nutrient intakes expressed as % of RDI:
		
		• Calories: 94%
		• Calcium: 44%
		• Iron: 196%
		• Vitamin A: 28%
		• Thiamin: 485%
		• Niacin: 126%
		• Vitamin C: 89%
			Anthropometric:
			• No major growth retardation or stunting reported
			
			Clinical:
			• Atrophic lingual papillae: 12%
			• Depigmentation of hair: 9% (95% of children <15 yr of age)
			• Follicular hyperkeratosis: 7%
			• Thyroid enlargement: 16%

Table 8 (continue). Investigations in Kenya.

Authors, year Survey area (*Ethnic group*) Sample (n)	Dietary intake	Dietary adequacy	Health status measures and prevalence of diseases

Korte *et al.*[112]
1969
Mwea-Tebere Villages (*Kikuyu*)
  ○ Mahigaine village
  ○ Kirogo village
 	
Total N = 251
	Breakfast:
	• Tea with sugar and sometimes milk, *uji** or leftovers from day before[++]
	
	Main foods:
	• Maize (57.3%) and legumes (24.5%)
	• Less frequently consumed: rice[+++] and other cereals, potatoes, meat, milk
	• Scarcely available: green vegetables
	• Children (up to 4-5 years) were given rice and English potatoes frequently
	
	Seasonal foods:
	• Higher consumption of rice, vegetable oil[++] and onion, and lower consumption of maize, legumes and sugar at the end of dry season (March)
		• Average calories: 2,604kcal
		• Fat: 9% of caloric intake
		• Protein score: 61.8/100
		• Carbohydrate score: 76.6/100
		• Calcium: 321mg/d
		• Iron: 27.5mg/d
		• Vitamin A: 308I.U./d
		• Vitamin B$_1$: 1.3mg/d
		• Vitamin B$_2$: 1.0mg/d
		• Vitamin C: 21mg/d
			Anthropometric:
			• Deviation towards lower than normal values
			
			Clinical:
			• High incidence of protein-energy malnutrition in children 1-4 yrs
			• Dyspigmentation (skin/hair): 44%
			• Atrophic lingual papillae: 21%
			• Xerosis of skin: 30%
			• Follicular hyperkeratosis: 12%
			• Corneal vascularization: 9.5%
			
			Parasitic infections (% affected)
			• *Ascaris lumbricoides*: 14.3%
			• *Shistosoma mansoni*: 7.6%
			
			Biochemical:
			• Anemia prevalent
			• Protein in serum within normal range
			• Retinol within normal range
			• Carotene within normal range

Table 8 (continue). Investigations in Kenya.

Authors, year Survey area (*Ethnic group*) Sample (n)	Dietary intake	Dietary adequacy	Health status measures and prevalence of diseases

Bohdal et al.[116]
1964 -1968
 	
Eastern Province
○ Nyaami village (*Wakamba*)
   survey 1: n = 18 families
   survey 2: n = 17 families
 	
Total N = 225
	Staple foods - survey 1
	• Millet flour (main cereal), sorghum, milk and salt
	
	Staple foods - survey 2#
	• Famine relief: yellow maize dry, yellow maize flour
	• Millet flour, milk and salt
	
	Breakfast
	• *Uji**
	Lunch
	• *Uji**
	Supper
	• *Uji**
	• Boiled maize
	• *Ugali*** and skim milk
	
	Notes:
	• The *uji** in this areas had milk added.
	• Higher millet and sorghum consumption in the first survey
		• Calories: 2265kcal/d
		• Protein: 70g/d
		• Calcium: 263mg/d
		• Iron: 17mg/d
		• Vitamin A: 383I.U./d
		• Vitamin B_1: 2.09mg/d
		• Vitamin B_2: 0.94mg/d
		• Niacin: 12.7mg/d
		• Vitamin C: 3.8mg/d
			Anthropometric:
			• *Children*: Signs of kwashiorkor or marasmic-type of muscle wasting
			
			Clinical:
			• Splenomegaly 18.7%
			• Hepatomegaly 26%
			• Pallor of membranes: 23%
			• Atrophic papillae: 45%
			• Angular palpebritis: 19.7%
			• Xerosis conjunctivae: 21%
			• Endemic *Visceral leishmaniasis* and *African trypanosomiasis*
			
			Biochemical:
			Anemia prevalent

RDI = Recommended dietary intake
*Thin soup made out of maize flour
**Stiff maize porridge, mainly made from maize
†Twelve different wild vegetables with high β-carotene were identified
‡Consumption of cheap vegetable oils and onions increased similarly to the rice consumption. They were used to improve the taste of disliked and leftover foods
‡‡Although rice was available in every household it was disliked due to the fact that after its consumption a feeling of hunger returned quickly. Rice consumption was higher when maize and legumes were not abundantly available
#Nyaani village was hit by drought and was under famine relief during this survey

Table 8 (continue). Investigations in Kenya.

Authors, year Survey area (*Ethnic group*) Sample (n)	Dietary intake	Dietary adequacy	Health status measures and prevalence of diseases
Bohdal et al.[116] 1964-1968 Nyanza Province o Masumbi village (*Luo*) survey 1: n = 116 survey 2: n = 97	*Staple foods – survey 1:* • Sorghum (main cereal), cassava four, green vegetables, onions, salt, fish (higher consumption of dry than fresh fish) *Staple foods - survey 2:* • White maize (main cereal), legumes, green vegetables, onions, salt, fish *Breakfast:* • Tea and *uji** *Lunch:* • *Ugali*** and green leaves *Supper:* • *Ugali*** and green leaves	Survey 1 Survey 2 • Calories: 2498kcal/d 2471kcal/d • Protein: 73g/d 86g/d • Calcium 987mg/d 534mg/d • Iron: 27mg/d 23mg/d • Vitamin A: 1305I.U./d 1212I.U./d • Vitamin B₁: 1.8mg/d 2mg/d • Vitamin B₂: 0.9mg/d 1.2 mg/d • Vitamin C: 87 mg/d 67 mg/d	Clinical: • Xerosis conjunctivae: 3.2% • Splenomegaly: 40% • Hepatomegaly: 28% *Parasitic infections:* 61% • *Ascaris lumbricoides*: 43.7% • African trypanosomiasis Biochemical examination • Anemia prevalent
Eastern Province o Uthiuni village (*Wakamba*) n = 133	*Staple foods:* • Yellow maize (main cereal from famine relief), beans, green leaves, wheat flour, cabbage, potato, tomato, fat, onions, fresh milk • Higher consumption of wheat flour than in other areas - used to make *uji** or chapatti eaten with vegetables and occasionally with meat stew *Breakfast:* • Tea and *uji** *Lunch:* • Maize and beans • *Ugali*** or *uji* with vegetable *Supper:* • Maize and bean mixture	• Calories: 1426kcal/d • Protein: 45g/d • Calcium: 250mg/d • Iron: 15mg/d • Vitamin A: 1179I.U./d • Vitamin B₁: 1.2mg/d • Vitamin B₂: 0.64mg/d • Niacin: 7.5mg/d • Vitamin C: 84 mg/d	Clinical: • Hepatomegaly: 59.6% • Splenomegaly: 17% • Dyspigmentation of skin: 49% • Thinness of hair: 44% • Lack of lustre: 29% • Pallor of membranes: 34% • Parotid enlargement: 24.6% Biomedical examination • Anaemia prevalent

Table 8 (continue). Investigations in Kenya.

Authors, year Survey area (*Ethnic group*) Sample (n)	Dietary intake	Dietary adequacy	Health status measures and prevalence of diseases

Bohdal et al.[116]
1964-1968
 	
Central Province
  o Ngamwa village (*Kikuyu*)
     n = 142
	Staple foods:
	• White maize flour (main cereal), beans, fresh milk, green maize, salt, green leaves, onions, bananas, cabbage, English potatoes, sugar, fat, tea
	Breakfast:
	• *Uji** or tea
	Lunch
	• Maize and bean mixture including green leaves
	• Potato-banana mixture
	Supper
	• Maize-bean mixture including green leaves
		• Calories: 1727kcal/d
		• Protein: 48g/d
		• Calcium: 259mg/d
		• Iron: 17mg/d
		• Vitamin A: 1538I.U./d
		• Vitamin B$_1$: 1.4mg/d
		• Vitamin B$_2$: 0.84mg/d
		• Niacin: 9.7mg/d
		• Vitamin C: 130mg/d
			Clinical:
			• Low incidence of protein deficiency (parotid enlargement 6.4%; thinness of hair 1.8%)
			• Thyroid enlargement: 24.7%
			• Atropic lingual papillae: 21%
			• Xerosis conjunctivae: 21%
			• Corneal vascularisation: 19.3%
			
			Parasitic infections: 40.7%
			• *Ascaris lumbricoides*: 25.8%

Nyanza Province
  o West Koguta village (*Luo*)
     survey 1: n = 149
     survey 2: n = 141
	Staple foods:
	• Sorghum (main cereal), milk, groundnuts, maize (dry), cassava flour
	Breakfast:
	• Tea and *uji** (equal popularity)
	Lunch:
	• *Ugali*** and green vegetables
	• *Ugali*** and fish (most popular)
	Supper:
	• *Ugali*** and fish
	• *Ugali*** and green vegetables
	
	Note: survey 2 revealed slightly reduced consumption of cereals and legumes but an increased consumption of fish in the sample.
		Survey 1 Survey 2
		• Calories: 1671kcal/d 1487kcal/d
		• Protein: 55g/d 52g/d
		• Calcium: 338mg/d 326mg/d
		• Iron: 15mg/d 13mg/d
		• Vitamin A: 5961.U./d 7041.U./d
		• Vitamin B$_1$: 1.47mg/d 1.14mg/d
		• Vitamin B$_2$: 0.68mg/d 0.65mg/d
		• Niacin: 13.6mg/d 8.6 mg/d
		• Vitamin C: 52mg/d 81mg/d
			Clinical:
			• Splenomegaly: 19%
			• Hepatomegaly: 9.8%
			• Spleno-hepatomegaly 6%
			• Xerosis conjunctivae: 9.8%
			• Thyroid enlargement: 8.6%
			
			Parasitic infections: 22.3%
			• *Schistosoma mansoni*: 1.1%
			
			Biochemical:
			• Anemia prevalent

Table 9. Investigations in Uganda.

Authors, year Survey area (*Ethnic group*) Sample (n)	Dietary intake	Dietary adequacy	Health status measures and prevalence of diseases

Jelliffe et al. [117]
1963
 	
Acholi district:
○ Rural area: Acholiland (*Acholi*)
   n = 222 (newborn infants)
   n = 384 (1-3 yr olds)
 	
○ Urban area: Kampala (*Acholi*)
   n=79 (newborn infants)
   n=53 (1-3 yr olds)
	Diet of rural Acholi children:
	• Breast-feeding until 2nd year of life
	• No bottle feeding existent
	• Soft and semi-solid foods introduced from second six months onwards
	• Main infant foods: eleusine millet, beans, green vegetables, sesame, meat
	
	Diet of urban Acholi children
	• Breast-feeding carried out until 16-18 months
	• Main infant foods: maize, potatoes (sweet or European), plantain, cassava, green vegetables, beans
		• Data not reported	Anthropometric (% affected):
			
			                                Infants 1-3 yr old
			Rural
			Normal weight range 27.9 42.2
			Third degree malnutrition 5.9 3.4
			Urban
			Normal weight range 69.9 58.5
			Third degree malnutrition 1.3 1.9
			
			Clinical (% affected):
			                                Infants 1-3 yr old
			Rural
			Kwashiorkor: 0.7 0.4
			Nutritional marasmus: 0.7 0.6
			Hypochromotrichia: 8.5 4.6
			Anemia: 41.9 22.5
			Palpable livers: 18.8 32.7
			Enlarged spleen: 55.1 63.5
			
			Urban 1-3 yr old
			• Kwashiorkor: 9.0
			• Hypochromotrichia: 22.8
			
			Parasitic (%) infections:
			                                Infants 1-3 yr old
			Rural
			• *Pl. Falciparum*: 53.2 53.3
			• *Pl. Malariae*: 23.4 27.5
			
			Biochemical:
			Haemoglobin (g%) (rural and urban combined)
			• Infants: 7.7
			• 1-3 yr olds: 8.6
			• 3% infants and 2% of 1-3 yr olds had hemoglobin < 5

3.6. References

1. Wild S, Roglic G, Green A, Sicree R, King H. Global prevalence of diabetes: Estimates for 2000 and projections for 2030. *Diabetes Care*. 2004; 27:1047-1053.
2. World Health Organization. Diet, physical activity and health. Fifty-five world health assembly A55/16. Geneva: World Health Organization; 27 March 2002:1-5.
3. Olshansky S, Passaro D, Hershow R, et al. A potential decline in life expectancy in the United States in the 21st century. *N Engl J Med*. 2002; 352(11):1138-1145.
4. Raschke V, Oltersdorf U, Elmadfa I, Wahlqvist M, Cheema B, Kouris-Blazos A. Content of a novel online collection of traditional East African food habits (1930s-1960s): Data collected by the *Max-Planck Nutrition Research Unit*, Bumbuli, Tanzania. *Asia Pac J Clin Nutr*. 2007; 16(1):140-151.
5. Dicko M, Hilhorst R, Gruppen H, et al. Comparison of content in phenolic compounds, polyphenol oxidase, and peroxidase in grains of fifty sorghum varieties from Burkina Faso. *J Agric Food Chem*. 2002; 50(13):3780-3788.
6. Nishizawa N, Shimanuki S, Fujihashi H, Watanabe H, Fudamoto Y, Nagasawa T. Proso millet protein elevates high plasma level of high-density lipoprotein: A new food function of proso millet. *Biomed Environ Sci*. 1996; 9:209-212.
7. Gooneratne J, Munasinghe L, Senevirathne W. Millet bran and corn bran lowers plasma total and LDL cholesterol levels in hypercholesterimic subjects. *Ann Nutr Metab*. 2005; 49(Suppl. 1):304.
8. Mossanda K, Kundu J, Na H, et al. Investigating African diet and

beverage in Africans with low risk of digestive and liver cancers. *Ann Nutr Metab.* 2005; 49(S1):168.

9. Hou W, Lee M, Chen H, et al. Antioxidant activities of dioscorin, the storage protein of yam (*Dioscorea batatas Decne*) tuber. *J Agric Food Chem.* October 2001; 49(10):4956-4960.

10. Mulokozi M, Hedren E, Svanberg U. In vitro accessibility and intake of beta-carotene from cooked green leafy vegetables and their estimated contribution to vitamin A requirements. *Plant Foods Hum Nutr.* 2004; 59(1):1-9.

11. Oboh G, Akindahunsi A. Change in the ascorbic acid, total phenol and antioxidant activity of sun-dried commonly consumed green leafy vegetables in Nigeria. *Nutr Health.* 2004; 18(1):29-36.

12. Johns T, Faubert G, Kokwaro J, Mahunnah R, Kimanani E. Anti-giardial activity of gastrointestinal remedies of the Luo of East Africa. *J Ethnopharmacol.* April 1995; 46(1):1-23.

13. Bangana A, Dossou N, Wade S, Guiro A, Lemonnier D. Cholesterol lowering effects of Okra (*Hibiscus esculentus*) in Senegalese adult men. *Ann Nutr Metab.* 2005; 49(S1):199.

14. Sankar D, Sambandam G, Ramakrishna Rao M, Pugalendi K. Modulation of blood pressure, lipid profiles and redox status in hypertensive patients taking different edible oils. *Clin Chim Acta.* May 2005; 355(1 2):97-104.

15. Sundram K, Sambanthamurth R, Tan Y. Palm fruit chemistry and nutrition. *Asia Pac J Clin Nutr.* 2003; 1(3):355-362.

16. Wattanapenpaiboon N, Wahlqvist M. Phytonutrient deficiency: The place of palm fruit. *Asia Pac J Clin Nutr.* 2003; 12(3):363-368.

17. Solomons N, Orozco M. Alleviation of vitamin A deficiency with palm fruit and its products. *Asia Pac J Clin Nutr.* 2003; 12(3):373-

384.

18. Li D, Siriamornpun S, Wahlqvist M, Mann N, Sinclair A. Lean meat and heart health. *Asia Pac J Clin Nutr.* 2005; 14(2):113-119.

19. Maletnlema T. A Tanzanian perspective on the nutrition transition and its implication for health. *Public Health Nutr.* 2002; 5(1A):163-168.

20. Nettleton J, Steffen L, Mayer-Davis E, et al. Dietary patterns are associated with biochemical markers of inflammation and endothelial activation in the Multi-Ethnic Study of Atherosclerosis (MESA). *Am J Clin Nutr.* June 2006; 83(6):1369-1379.

21. Cannon G. Nutrition: The new world disorder. *Asia Pac J Clin Nutr.* 2002; 11 (Suppl):498-509.

22. Popkin B. Global nutrition dynamics: The world is shifting rapidly toward a diet linked with noncommunicable diseases. *Am J Clin Nutr.* Aug 2006; 84(2):289-298.

23. Hawkes C. The role of foreign direct investment in the nutrition transition. *Public Health Nutr.* June 2005; 8(4):357-365.

24. Hawkes C. Uneven dietary development: Linking the policies and processes of globalization with the nutrition transition, obesity and diet-related chronic diseases. *Global Health.* March 2006; 2(4):1-18.

25. Chopra M, Darnton-Hill I. Responding to the crisis in sub-Saharan Africa: The role of nutrition. *Public Health Nutr.* August 2006; 9(5):544 550.

26. Lang T. Diet, health and globalization: Five key questions. *Proc Nutr Soc.* 1999; 58(2):335-343.

27. Chossudovsky M. *The Globalization of Poverty and the New World Order.* Vol 2. Quebec, Canada: Global Outlook; 2003:1-376.

28. Ayittey G. *Africa Unchanged*: The Blueprint for Africa's Future.

New York: Palgrave Macmillan; 2005:1-480.

29. Tarver J. *The Demography of Africa*. Westport, CT: Praeger; 1996: 1-288.

30. Popkin BM. Part II. What is unique about the experience in low- and middle-income less-industrialized countries compared with very-high-income industrialized countries? The shift in stages of the nutrition transition in the developing world differs from past experiences! *Public Health Nutr.* 2002; 5(1A):205-214.

31. Voster H, Bourne L, Venter C, Oosthuizen W. Contribution of nutrition to the health transition in developing countries: A framework for research and intervention. *Nutr Rev.* 1999; 57(11):341-349.

32. Shiva V. *Stolen Harvest*. London: Zed Books; 2000: 1-146.

33. Pringle P. *Food Inc.*: Mendel to Monsanto - The Promises and Perils of the Biotech Harvest. New York: Simon and Schuster; 2005: 1-229.

34. United Nations Conference on Trade and Development (UNCTAD). World Investment Report. FDI policies for development: National and international perspectives. Geneva: UNCTAD; September 2003: 1-322.

35. United Nations Conference on Trade and Development (UNCTAD). World Investment Directory. UNCTAD. Available at: http://www.unctad.org/Templates/Page.asp?intItemID=3198&lang=1 (Accessed September 2006).

36. Yamori Y, Miura A, Taira K. Implications from and for food cultures for cardiovascular diseases: Japanese food, particularly Okinawan diets. *Asia Pac J Clin Nutr.* 2001; 10(2):144-145.

37. Cockerham W, Yamori Y. Okinawa: An exception to the social

gradient of life expectancy in Japan. *Asia Pac J Clin Nutr.* June 2001; 10(2):154-158.

38. Trichopoulou A, Critselis E. Mediterranean diet and longevity. *Eur J Cancer Prev.* October 2004; 13(5):453-456.

39. Serra-Majem L, Roman B, Estruch R. Scientific evidence of interventions using the Mediterranean diet: A systematic review. *Nutr Rev.* February 2006; 64(Suppl 1): 27-47.

40. MacLennan R, Zhang A. Cuisine: The concept and its health and nutrition implications - global.*Asia Pac J Clin Nutr.* 2004; 13(2):131-135.

41. Liu X, Li Y. Epidemiological and nutritional research on prevention of cardiovascular disease in China. *Br J Nutr.* December 2000; 84(Suppl 2):199-203.

42. Fallon S. Ancient dietary wisdom for tomorrow's children. The Weston A. Price Foundation. Available at: http://www.westonaprice.org/traditional_diets/ancient_dietary_wisdom.html (Accessed September 2006).

43. Price W. *Nutrition and Physical Degeneration.* 14th edition. La Mesa, CA: Price-Pottenger Foundation; 2000:513.

44. Raschke V. East African food habits on-line. In: Wahlqvist ML. HEC Press. Available at: http://www.healthyeatingclub.org/Africa/index.htm.

45. Latham M. Nutritional studies in Tanzania (Tanganyika). *J Am Diet Assoc.* 1967; 7:31-72.

46. Laurie W, Brass W, Sewell P, et al. A health survey in Bukoba district, Tanganyika: East African High Commission; 1951:1-138.

47. Latham M. Nutritional problems in Tanganyika. Report of 6th International Congress of Nutrition, Edinburgh: Livingstone;

1964:449-455.

48. Zanzibar Protectorate. Nutritional review of the natives of Zanzibar. Zanzibar: Zanzibar Protectorate; 1937:1-20.

49. Holmes E, Stanier M, Thompso M. The serum protein pattern of Africans in Uganda: Relation to diet and malaria. *Trans Roy Soc Trop Med Hyg.* 1955; 49:376-384.

50. Harkness J. Deficiency diseases in the Bukoba district, Tanganyika territory. *Trans R Soc Trop Med Hyg.* January 1935; 28(4):407-410.

51. Loewenthal J. An inquiry into vitamin A deficiency among the population of Teso, Uganda, with special reference to school children. *Ann Trop Med.* 1935; 29:349-360.

52. Kenneth M. Report on dietary surveys and nutritional assessment in East Africa. Nairobi: East African Bureau of Research in Medicine and Hygiene; 1953:1-18.

53. Leonard P. Serum and liver levels of Vitamin A in Ugandans. *East Afr Med J.* April 1964; 41(4):133-136.

54. Uganda Protectorate Nutrition Committee. Review of nutrition in Uganda. Entebbe: Nutrition Committee, Government Printer Uganda; 1945:1-19.

55. Dean R. Protein supply in Uganda. *East Afr Med J.* July 1962; 39:493-500.

56. Jelliffe D, Jelliffe E, Benett F, White R. Ecology of childhood disease in the Karamojong in Uganda. *Arch Environ Health.* 1964; 9:25-36.

57. Loewenthal L. Interim peport on nutrition in Uganda. Entebbe: Government Printer, Uganda; 1940:1-7.

58. Leonard P. Vitamin E deficiency in Uganda African subjects. *Trans R Soc Trop Med Hyg.* November 1964; 58(6):517-520.

59. Burgess H. Protein-calorie malnutrition in Uganda. I. General. *East Afr Med J.* July 1962; 39:357-361
60. Welbourn HF. Notes on differences between Baganda and Luo children in Kampala. *East Afr Med J.* July 1955; 32:291-298.
61. Burgess AP. Calories and proteins available from local sources for Uganda Africans in 1958 and 1959. *East Afr Med J.* 1962; 39:449-465.
62. Shaper A, Jones M, Kyobe J. Plasma-lipids in an African tribe living on a diet of milk and meat. *Lancet.* 1961; 162(2):1324-1327.
63. Gilks J. Dietetic problems in East Africa. *East Afr Med J.* 1933; 10:254-265.
64. Levy A. Hemoglobin differences among Kenyan tribes. *Am J Trop Med Hyg.* 1969; 18(No. 1):138-146.
65. Worthington E. On the food and nutrition of African natives. *Afr.* 1936; 9:150-165.
66. Price W. Field studies among some African tribes on the relation of the nutrition to the incidence of dental caries and dental arch deformities. *J Am Dent Assoc.* May 1936; 23:876-890.
67. Moller M. African child health: Custom, pregnancy and child rearing in Tanganyika. *J Trop Pediatr.* 1961; 7:66-80.
68. Lema N. Tribal customs in infant feeding II among the Chagga. *East Afr Med J.* 1963; 40:370-375.
69. Nguma G. Food science and applied human nutrition research: Progress report of a dietary survey in Kilosa district: Research and Training Institute Ilonga Food Science Section; 1969:11.
70. Jerrard R. Tribes of Tanganyika, their districts, usual dietary and pursuits. Dar es Salaam: Government Printer; 1935:1-24.
71. Tanner R. A preliminary enquiry into the Sukuma diet in the Lake

Province, Tanganyika territory. *East Afr Med J.* August 1956; 33(8):305-324.

72. Gerlach LP. Economy and protein malnutrition among the Digo. Anthropology No. 29. Minnesota: The Minnesota Academy of Science; 1961:3-13.

73. Culwick A, Culwick G. A study of factors governing the food supply in Ulanga, Tanganyika territory. *East Afr Med J.* 1939; 16:43-61.

74. Rutishauser I. Custom and child health in Buganda. *Trop Geogr Med.* 1963; 15:138-147.

75. Schaffer R, Finklestein F. The food and growth of Gogo children. Stenciled Paper. 1963:1-10.

76. Jelliffe D, Bennett J. Cultural and anthropological factors in infant and maternal nutrition. *Fed Proc.* 1961; 20:185-187.

77. Smith W, Smith E. Native diet of Zanzibar. *East Afr Med J.* November 1935; 11(8):246-251.

78. Rwegelera GG. Tribal custom in infant feeding I. among the Haya. *East Afr Med J.* July 1963; 40(7):366-369.

79. Trant H. Food taboos in East Africa. *Lancet.* 1954; 2:703-705.

80. Cremer HD. *Verbesserung der Ernährungssituation in Ostafrika.* Vol. Stuttgart: Ernst K lett Verlag; 1966:1-152.

81. Callanan J. Notes on the foodstuffs of the Luo tribe. *Ken Med J.* 1926; 3:58-60.

82. Keller R. *Studie zur Ernährung bei zwei Stämmen in Nord-Tanganyika.* Köln, West Germany: Westdeutscher Verlag; 1965:1-49.

83. Gower R. The effect of a change of diet on Masai schoolboys. Notes of the investigation into the health of Masai schoolboys at Monduli School during 1945. 1948:17-19.

84. Schwartz R, Dean R. An investigation of the daily intakes of food of individual boys at a boarding school in Uganda. *Br J Nutr.* 1955; 9:230.
85. Allen K. The monotonous diet of the African. *East Afr Med J.* March 1955; 32:95-97.
86. Welbourn H. The danger period during weaning (Part II). *J Trop Pediatr.* September 1955; 1:98-105.
87. Schlage C. *Analysis of Some Important Foodstuffs of Usambara.* In: Kraut H, Cremer HD, Ifo Institut für Wirtschaftsforschung e.V. M, eds. Investigations into Health and Nutrition in East Africa, Afrikastudien. München: Weltforum Verlag, München; 1969:85-178.
88. Raymond W, Jojo W, Nicodemus Z. The nutritive value of some Tanganyika foods II - Cassava. *East Afr Med J.* 1941; 6:154-159.
89. Naik A. Chemical composition of Tanzania feedstuffs. *East Afr Agric For J.* 1969; 32(2):201-205.
90. Muskat E, Keller W. Nutzung von Wildpflanzen als zusätzliche Vitaminquelle der Ernährung der Bevölkerung Kenias. *Internat Z Vit Forschung.* September 1968; 38:538-544.
91. French MH. Some notes on the common foodstuffs used in the diet of East African natives. *East Afr Med J.* 1937; 13:374-378.
92. Harvey D. The chemical composition of some Kenya foodstuffs. Records of the medical research laboratory. Colony and protectorate of Kenya, Medical Department. 1967: 201-205.
93. Culwick A, Culwick G. Nutrition and native agriculture in East Africa. *East Afr Med J.* 1941; 6:146-153.
94. Farnsworth A. The diet of the African Soldier. *East Afr Agric For J.* 1943; 20(7):207-213.

95. Orr J, Gilks J. The physique and health of two African tribes. Special Report Series Medical Research Council. London: His Majesty's Stationery Office; 1931:1-82.
96. Schlage C. *Observation at the Local Market in Usambara*. In: Kraut H, Cremer HD, Ifo Institut für Wirtschaftsforschung e.V. M, eds. Investigations into health and nutrition in East Africa, Afrika Studien. München: Weltforum Verlag, München; 1969:72-84.
97. Procter R. The Kikuyu market and Kikuyu diet. *Ken Med J.* 1926; 3:15-22.
98. Raymond W. Reasons for a nutritional policy in Tanganyika. Dar es Salaam: Tanganyika Territory Med. Dep. Pamphlet. Government Printer. 1941; 35.
99. Rowland HAK. Anaemia in Dar es Salaam and methods for its investigation. *Trans R Soc Trop Med Hyg.* 1966; 60:143.
100. Gale GW. Food balance sheets for the African populations of East Africa. *East Afr Med J.* May 1960; 37(5):410-417.
101. Trant H. A report of a dietary survey in the Taveta-Pare area of Tanganyika. East African Medical Survey Annual Report: East African Institute; 1956:1-22.
102. Raymond W. Minimum dietary standards for East African natives. *East Afr Med J.* 1940; 17:249.
103. Gerlach L. Socio-cultural factors affecting the diet of the northeast coastal Bantu. *J Am Diet Assoc.* 1964; 45:420-424.
104. Philip C. Nutrition in Kenya: Notes on the state of nutrition on African children. *East Afr Med J.* 1943; 20:227-234.
105. Tanzania National Nutrition Unit. Report of a dietary survey in the Tabora region. Tanzania Nutrition Committee Report: Series 4a, The Tanzania National Nutrition Unit; 1967:1-11.

106. East African High Commission. Nutrition survey in the Kilimanjaro area: Government Printer; 1968:1-13.
107. Mann GV, Schaeffer RS, Anderson RS, Sandstead HH. Cardiovascular disease in the Masai. *J Atheroscler Res.* 1964; 4:289-312.
108. Courcy-Ireland M, Hosking H, Löwenthal L. An investigation into health and agriculture in Teso, Uganda. Nutrition Report No. 1. Teso: Agricultural Survey Committee; 1937:1-28.
109. McLaren D. Nutrition and eye disease in East Africa: Experience in Lake and Central Provinces, Tanganyika. *J Trop Med Hyg.* 1960; 63:101-122.
110. Kasper H, Schaefer HJ, Theermann P. *A Contribution to Determine the Nutritional State of the People Living in the Central Province of Kenya.* In: Kraut H, Cremer HD, Ifo Institut für Wirtschaftsforschung e.V. M, eds. Investigations into health and nutrition in East Africa, Afrika Studien. München: Weltforum Verlag, München; 1969:227-240.
111. Trowell H. Calorie and protein requirements of adult male Africans. *East Afr Med J.* May 1955; 32(5):153-163.
112. Korte R. *The Nutritional and Health Status of the People Living on the Mwea-Tebere Irrigation Settlement.* In: Kraut H, Cremer HD, Ifo Institut für Wirtschaftsforschung e.V. M, eds. Investigations into health and nutrition in East Africa, Afrika Studien. München: Weltforum Verlag, München; 1969:267-331.
113. Kreysler J, Schlage C. *The Nutrition Situation in the Pangani Basin.* In: Kraut H, Cremer HD, Ifo Institut für Wirtschaftsforschung e.V. M, eds. Investigations into health and nutrition in East Africa. München: Weltforum Verlag, München; 1969:85-178.

114. Popleau W, Schlage C, Keller W. *Nutrition and Health in Usambara*. In: Kraut H, Cremer HD, Ifo Institut für Wirtschaftsforschung e.V. M, eds. Investigations into health and nutrition in East Africa, Afrika Studien. München: Weltforum Verlag, München; 1969:25-45.

115. Keller W, Muskat E, Vlader E. *Some Observations Regarding Economy, Diet and Nutritional Status of Kikuyu Farmers in Kenya*. In: Kraut H, Cremer HD, Ifo Institut für Wirtschaftsforschung e.V. M, eds.Investigations into health and nutrition in East Africa, Afrika Studien. München: Weltforum Verlag, München; 1969:241-266.

116. Bohdal M, Gibbs N, Simmons W. Nutrition surveys and campaigns against malnutrition in Kenya. Report to the Ministry of Health of Kenya. Geneva, Switzerland: WHO/FAO/UNICEF; 1969:1-171.

117. Jelliffe DB, Bennett JF, Stroud CE, Welbourn HF, Williams MC, Jelliffe EFP. The health of Acholi children. *Trop Geogr Med.* 1963; 15 (Community and child health studies in East Africa No. 5):411-421.

118. Kraut H, Cremer HD, Attems MG, et al. *Investigations into Health and Nutrition in East Africa*. Afrika Studien. München: Weltforum Verlag, München; 1976:1-342.

119. Maundu P, Imbumi M. East Africa. In: Katz S, Weaver W, eds. *Encyclopedia of Food and Culture.* Vol. 3 New York: Thomson and Gale; 2003:27-34.

120. Maundu P, Ngugi G, Kabuye C. *Traditional Food Plants of Kenya.* Nairobi: Kenrik; 1999:1-269.

121. Read M. Native standards of living and African culture change. *Africa.* 1938; 11(Suppl. 3):1-64.

122. Robson J. Changing food habits in developing countries. *Ecol Food*

Nutr. 1976; 4:251-256.

123. Hughes C, Hunter J. *The Role of Technological Development in Promoting Disease in Africa.* In: Farvar M, Milton J, eds. The careless technology: Ecology and international development. New York: The Natural History Press; 1972:69-101.

124. Tabuti J, Dhillion S, Lye K. The status of wild food plants in Bulamogi County, Uganda. *Int J Food Sci Nutr.* September 2004; 55(6):485-498.

125. Food and Agriculture Organization of the United Nations (FAO). *Traditional Food Plants*: A resource book for promoting the exploitation and consumption of food plants in arid, semi-arid and sub-humid lands of Eastern Africa. Food and Nutrition Paper 4-2; 1988.

126. Gura S. A note on traditional food plants in East Africa: Their value for nutrition and agriculture. *Food Nutr.* 1986; 12(1):18-22.

127. Okigbo B. Broadening the food base in Africa: the potential of traditional food plants. *Food Nutr.* 1986; 12(1):4-17.

128. Demment M, Young M, Sensenig R. Providing micronutrients through food-based solutions: A key to human and national development. *J Nutr.* November 2003; 133(11 Suppl 2):3879-3885.

129. Welch R, Graham R. A new paradigm for world agriculture: Meeting human needs productive, sustainable and nutritious. *Field Crop Res.* 1999; 60:1-10.

130. Kuhnlein H, Receveur O. Dietary change and traditional food systems of indigenous people. *Annu Rev Nutr.* 1996; 16:417-442.

131. Johns T, Eyzaguirre P. Linking biodiversity, diet and health in policy and practice. *Proc Nutr Soc.* May 2006; 65(2):182-189.

132. FAO. Financing normal levels of commercial imports of basic

foodstuffs. Rome: Food and Agriculture Organization of the United Nations; 2003:1-135.

133. Cannon G. Nutrition: The new world map. *Asia Pac J Clin Nutr.* 2002; 11(Suppl):480-497.

134. Poulter N, Khaw K, Hopwood B, et al. Blood pressure and its correlates in an African tribe in urban and rural environments. *J Epidemiol Community Health.* 1984; 38(3):181-185.

135. Poulter N, Khaw K, Hopwood B, et al. Blood pressure and associated factors in a rural Kenyan community. *J Hypertens.* 1984; 6:810-813.

136. Poulter N, Khaw K, Hopwood B, et al. The Kenyan Luo migration study: observations on the initiation of a rise in blood pressure. *BMJ.* April 1990; 300(6730):967-972.

137. Bourne LT, Lambert EV, Steyn K. Where does the black population of South Africa stand on the nutrition transition? *Public Health Nutr.* 2002; 5(1A):157-162.

138. Njelekela M, Ikeda K, Mtabaji J, Yamori Y. Dietary habits, plasma polyunsaturated fatty acids and selected coronary disease risk factors in Tanzania. *East Afr Med J.* Nov. 2005; 82(11):572-578.

139. Alexandratos. *Costs and Implications for a Sustainable Utilization of Plant Genetic Resources for Food and Agriculture.* In: Virchow D, ed. Conservation of Genetic Resources: Springer; 1999.

140. Popkin B, Bisgrove E. Urbanization and nutrition in low-income countries. *Food Nutr Bull.* 1988; 10(1):3-23.

141. Oniang'o RK. Food habits in Kenya: The effects of change and attendant methodological problems. *Appetite.* 1999; 32:93-96.

142. Maundu P. The status of traditional vegetable utilization in Kenya. International Plant Genetic Resource Institute (IPGRI). Available at:

http://www.ipgri.cgiar.org/publications/HTMLPublications/500/ch09.htm (Accessed May 2006).

143. Rubaihayo E. Conservation and use of traditional vegetables in Uganda. International Plant Genetic Resources Institute (IPGRI). Available at: http://www.ipgri.cgiar.org/publications/HTMLPublications/500/ch15.htm (Accessed May 2006).

144. Mazengo M, Simell O, Lukmanji Z, Shirima R, Karvetti R. Food consumption in rural and urban Tanzania. *Acta Trop.* December 1997; 68(3):313-326.

145. Popkin B, Lu B, Zhai F. Understanding the nutrition transition: Measuring rapid dietary changes in transitional countries. *Public Health Nutr.* 2002; 5(6A):947-993.

146. Allen L. Interventions for micronutrient deficiency control in developing countries: Past, present and future. *J Nutr.* November 2003; 133(11 Suppl 2):3875-3878.

147. Ramakrishnan U. Prevalence of micronutrient malnutrition worldwide. *Nutr Rev.* 2002; 60:36-52.

148. Allen L, Gillespie S. What works? A review of the efficacy and effectiveness of nutritional interventions. Geneva: ACN/SCN and Asian Development Bank; 2001. Available at: http://www.unsystem.org/SCN/archives/npp19/index.htm

149. World Health Organization. Global prevalence of vitamin A deficiency. Micronutrient Deficiency Information System, No. 2 (WHO/NUT/95.3) Geneva: WHO Division of Nutrition; 1995:1-126.

150. Sommer A, West K. *Vitamin A deficiency: Health, Survival and Vision.* Oxford: University Press; 1996.

15. UNICEF. *Vitamin A Global Initiative.* New York: UNIFEC; 1998.

152. Hewitt de Alcantara C. Modernizing Mexican agriculture: Socioeconomic implications of technological change, 1940-1970. Geneva: United Nations *Res Inst Soc Dev.*; 1976.
153. Cervinskas J, Lotf M. *Vitamin A Deficiency: Key resources in its prevention and elimination.* Ottawa: The Micronutrient Initiative.1996; Available at: http://www.micronutrient.org/resources/publications/pub19.htm
154. Unwin N, Mugusi F, Aspray T, et al. Tackling the emerging pandemic of non-communicable diseases in sub-Saharan Africa: The essential NCD health intervention project. *Public Health.* May 1999; 113(3):141-146.
155. FAO/ILSI. Preventing micronutrient malnutrition: A guide to food-based approaches. Washington, DC: International Life Sciences Institute/ILSI; 1997.
156. Murphy S, Allen L. Nutritional Importance of Animal Source Foods. *J. Nutr.* November 2003; 133:3932-3935.
157. Wahlqvist M. *Food and Nutrition.* 2nd edition; Crows Nest, NSW, Australia: Allen and Unwin; 2002.
158. Zagre N, Delpeuch F, Traissac P, Delisle H. Red palm oil as a source of vitamin A for mothers and children: Impact of a pilot project in Burkina Faso. *Public Health Nutr.* 2003; 6(8):733-742.
159. West C, Temalilwa C. *The Composition of Foods commonly Eaten in East Africa.* Wageningen: Wageningen Agricultural University; 1988.
160. Chweya J, Eyzaguirre P. *The Biodiversity of Traditional Leafy Vegetables.* Rome: International Plant Genetics Resource Institute (IPGRI); 1999.
161. Uiso F, Johns T. Consumption patterns and nutritional contribution

of *Crotalaria brevidens* (*mitoo*) in Tarime District, Tanzania. *Ecol Food Nutr.* 1996; 35(1):59-69.

162. Nyakeriga A, Troye-Blomberg M, Chemtai A, Marsh K, Williams T. Malaria and nutritional status in children living on the coast of Kenya. *Am J Clin Nutr.* 2004; 80(6):1604-1610.

4. COLONIAL AND NEOCOLONIAL FORCES AND THE ERADICATION OF TRADITIONAL FOOD HABITS IN EAST AFRICA: HISTORICAL PERSPECTIVE ON THE *NUTRITION TRANSITION*

Since the 1500, when the imperial powers of Europe sought to expand their empires through the colonization of sub-Saharan Africa, the effort has been to destroy the indigenous people, their way of life and their ancient knowledge, including a vast and incredible knowledge of food habits and their associated benefits to health and longevity.[1] Over the past few centuries, the means used to subjugate the population of Africa has shifted from overt force (e.g. genocide, slavery, seizure of arable land and resources)[2] to the implementation of a *neocolonial,* political-economic structure designed to oppress the African population through the creation of economic dependence (i.e. debt and taxes) and economic exploitation (by trade-policy reforms and transnational corporations).[3-6]

The inhumane impact of these *colonial* and *neocolonial* forces persists in sub-Saharan Africa today, largely unresolved, and becomes glaringly evident with investigation of the root causes of disease epidemics currently plaguing the indigenous people of this vast continent.[7] Over the past several decades, sub-Saharan Africa has experienced a rapid upsurge of non-communicable diseases (NCDs), which includes epidemics of obesity, diabetes, cardiovascular disease (CVD), and various cancers.[7] Within the next 20 years, sub-Saharan Africa can expect a three-fold increase in deaths due to CVD and a near three-fold increase in the incidence of type 2 diabetes.[8] Further, NCDs have not simply replaced infectious and malnutrition-related diseases throughout sub-Saharan Africa, they co-exist alongside classic nutritional deficiencies, famine, and infectious diseases resulting in a polarized and protracted *double burden* of disease.[9-12] By 2020, it is predicted that NCDs will account for 80% of the global disease burden, and will cause 70% of deaths in developing countries.[7] The health care systems in sub-Saharan Africa are either non-existent or are grossly inadequate to deal with this burgeoning *double burden* of disease and its myriad consequences (i.e. for healthcare, for economic viability and political stability).[13-15]

The NCD epidemics currently sweeping sub-Saharan Africa have been directly attributed to the *nutrition transition*, whereby traditional foods and food habits have been progressively replaced by the *globalized* food culture of the multinational corporations.[10, 16, 17] The transition to a simplified diet includes increased consumption of energy-dense food, which is high in saturated fat and low in unrefined carbohydrates.[18] These newly formed dietary patterns have been recognized as a marked contributor to NCD epidemics.[19] By contrast, we have presented scientific evidence that traditional East Africa food habits, including dietary intake of a broad spectrum of cereals, roots, tubers, fruits, vegetables, spices, fats, fish and wild bush meats, are associated with myriad health benefits according to the latest empirical investigations.[20]

Paradoxically, while the *globalized* food culture exerts a pathological effect[21-23] and traditional foods exert a protective effect on NCDs,[24-31] it is the *globalized* food habits that continue to be promoted by multinational cooperations, whilst scientists and politicians alike continue to affirm support for the 'battle" against obesity and diabetes.

Development of today's *globalized* food system, a system inherently connected to the global epidemics of NCDs, is rooted in the creation of policies and institutions which govern the production, trade, distribution and marketing of food.[32, 33] Currently, a handful of transnational corporations control this system and, as such, exert direct control over the creation of the latest NCD epidemics. The control of the transnational corporations is exerted by *scarcity-resulting from-abundance*, a practice resulting in the decrease of quality whole food options (*scarcity*) and the widespread distribution of insidious, low-quality processed foods (*abundance*).[20] This corporate philosophy is summarized succinctly by Mr. Ray Kroc, the founder of the McDonalds chain, as cited in Eric Schlosser's *Fast-Food Nation*:[34]

> *We have found out... that we cannot trust some people who are nonconformists... We will make conformists out of them... The organization cannot trust the individual; the individual must trust the organization.*

Throughout history, external influences have brought about changes in African food habits.[35-37] In centuries past, the overall objective of the colonial powers was to subjugate the population through the seizure of arable land and by controlling the distribution of food. Control of the food system (and hence health status and NCDs) is intimately linked to control of the population. The need to subjugate the population is inherent to the needs of a *New World Order*, an agenda for global hegemony driven by the transnational corporations and their political allies, as exposed by countless authors.[3, 4, 38] Indeed, the *globalized* food system has recently been described as *a weapon of control*.[39] Susan George[40] revealed this perspective over 30 years ago when she stated:

> *This is what food has become: a source of profits, a tool of economic and political control; a means of ensuring effective*

dominance over the world at large and especially over the 'wretched of the earth'.

While the simplification of the East African cuisine may be most apparent today, the *nutrition transition* has actually taken place over the past 400 years.[35-37, 41] Currently, there is an imperative need to investigate and disseminate information related to the factors historically responsible for the *nutrition transition* and its myriad repercussions in sub-Saharan Africa. Such inquiry is necessary to fully comprehend current NCD epidemics, and improve the health status of the marginalized indigenous people throughout the region. The abatement of NCD epidemics may involve the resurrection of the ancient, indigenous knowledge about traditional food habits.[20]

The purpose of the present article is to discuss factors which have underpinned the *nutrition transition* in the countries of East Africa, including Kenya, Uganda and Tanzania, from early colonization and the arrival of *Columbian Exchange* (i.e. Enormous widespread exchange of new and different goods from the Eastern and Western hemispheres that occurred after 1492) to the current oppressive, political-economic structure. A conceptual framework which outlines the *colonial* and *neocolonial* contributors to the eradication of traditional food habits throughout East Africa over the past few centuries has been provided in Figure 3 and will be discussed herein.

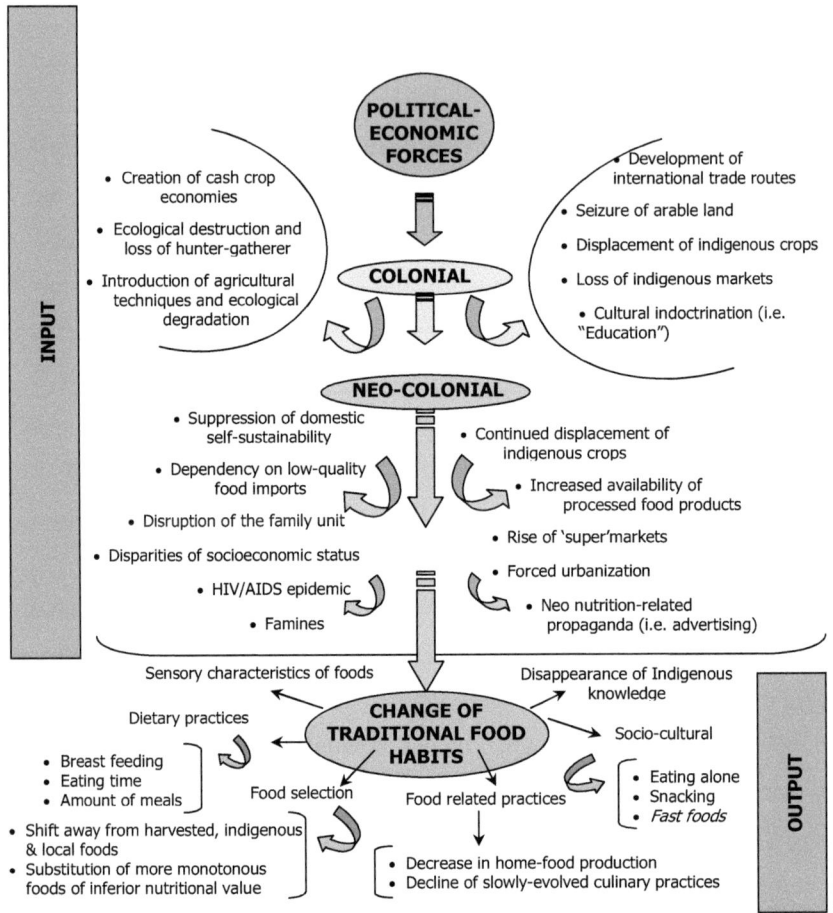

Figure 3. *Colonial* and *Neocolonial* contributors to the change in traditional East African food habits.

4.1. Colonial impact on the nutrition transition in East Africa

About five thousand years ago, much of East Africa was occupied by hunter-gatherers commonly referred to as *ndorobo*.[41] Although few of these people still exist, most of these groups were assimilated by later migrants and therefore lost their identity including their food culture. Our ancestors hunted big and small game and gathered wild foods such as fruits, nuts, tubers, honey, grasshoppers, caterpillars, termites, eggs, and some birds.[41] Today the contribution of gathering is less significant but many aspects of it remained. Today, agriculture is, by far, the most important production system in East Africa. Agriculture in East Africa was pioneered by *Cushitic* speakers from the Ethiopian highlands. Other cultivators came in from the south, west (*Bantu*), and northwest (*Nilotes*). The earliest food crops of agriculturalists in this region included sorghum, finger and pearl millets, hyacinth (*lablab*) beans, bambara groundnuts, bottle gourds, cowpeas and yams.[41-43] These staple foods have been associated with numerous health benefits.[20] Cultivated and wild vegetables, especially, wild green leaves, including amaranth, black nightshade, and red sorrel were important ingredients for use in sauces accompanying carbohydrate staples.[41]

4.1.1. Development of international trade routes

Food habits in East Africa began to shift for the first time in modern history perhaps in the 1400s with the development of the coastal trading towns, the creation of international trade routes[44] and the onset of colonial occupation.[2,3,5] Trade has spread foods around the world and has transformed food preferences, tastes and habits across all geographical regions and cultures.[45, 46] Two distinct events in relation to international trade had a profound influence on the food habits of the East African population. The first was the discovery and use of the sea route to India and Southeast Asia in the late 15th and early 16th century, and the second was the development of the international *Columbian Exchange* system which occurred with "discovery" of the Americas by Christopher Columbus in 1492.[44]

Through trade with Asia, East African farmers acquired a number of crops, including banana, plantain, cocoyam, and sugar cane, which were rapidly assimilated into the local diets.[41] *Columbian Exchange* also led to the introduction of staple crops from the Americas, including, most notably: maize, peanuts (groundnuts), potatoes, kidney beans, pumpkins, cassava (manioc), European cabbage, and kale (*Sumuka wiki*).[41, 47] These particular foods proved to be ecologically sustainable and thus rapidly altered and diversified food intake of East Africans. The use of these new introduced foods became widespread during the colonial period, from the 1850s to 1960s,[41] and rural inhabitants commonly replaced indigenous crops with overseas varieties. However, these introduced staple crops began to threaten the future of the robust indigenous crops including varieties of millet and sorghum.[48]

The predominance of introduced exotic foods in the East African diet was reflected in our recent review of the *Oltersdorf Collection*, a compilation of investigations on traditional East African foods and food habits from the 1930s-1960s, collected by the *Max-Planck Nutrition Research Unit* in Bumbuli, Tanzania.[20] Today, common staple dishes in Kenya includ, for example, *ugali* (a starchy staple food primarily made from maize), *githeri* (a mixture of maize and pulses - seeds of legumes, such as chickpeas, lentils, field peas, and peanuts), *pilau* (spiced rice cooked with meat), and chapatti (flat bread made of wheat flour). In Tanzania, current staple dishes includ *wali/pilau* and *makande* (both mixtures of maize and beans) and in Uganda, today's common staple dishes includ steamed *matooke* (banana), sweet potato and cassava staple dishes served with a groundnut sauce.[41]

4.1.2. Seizure of arable land

Land pressures in Kenya, Tanzania and Uganda during colonization fundamentally arose from colonial policies which enabled European settlers to seize control of so-called empty lands.[49, 50] These empty lands were in fact some of the most arable lands in East Africa.[49, 50] Driven from their lands, and denied equitable access to the natural resources, indigenous farmers were forced to congregate on small marginal plots.[49-51] Generally, the former indigenous land owners remained landless.[49-51]

The seizure of the arable land by the European colonizers (i.e. Portugal, Great Britain and Germany) was concomitant with a shift toward farming higher-yielding, less labor-intensive crops, such as maize and cassava. This shift caused joblessness for farmers/farm labourers and removed nutrient-dense indigenous cereals including millets and sorghums from the hitherto diverse diet.[52] Seizure of arable land also resulted in reduced livestock ownership among the indigenous people.[51] The loss of livestock numbers drastically reduced animal protein intake.[51, 53, 54] Inadequate protein intake, resulting in a high prevalence of protein-energy malnutrition, amongst the local, indigenous people of East Africa from the 1930s to 1960s has been well documented by the *Max-Plank Nutrition Research Unit*.[55-63]

4.1.3. Creation of cash crops economies

From the end of the 19th century, the colonial enterprise in East Africa aimed to advance a trading system which predominantly served European industries and consumers.[64] Primarily, it was the imperial power of Great Britain that controlled the production and export of East African-grown staple crops and other valuable commodities.[3] As part of this exploitation, the 1950s witnessed the emergence of cash crop economies.[65]

These cash crop economies were primarily based on the production of coffee, copra, cotton, sesame, peanuts, and sugar for the western markets of Europ, North America, and Australia. This cash crop production largely persists to the present day.[6, 66] Rural communities from the outset, were encouraged by colonialists to grow food crops for export in order to earn money to "improve their standards of living".[65] As a result, domestic food production became neglected by East Africans as they were forced to pay taxes to the imperial powers.[67]

4.1.4. Ecological destruction and the loss of hunter-gatherer areas

Diverse forest ecosystems were cleared for land needed to support cash-crop farming.[65] This removal of native flora eliminated many indigenous foods from the diet.[68] The consumption of several indigenous fruit species and wild food plants diminished in the traditional diets, affecting the taste and nutrient content of common dishes.[43, 69, 70] Utilization of wild bush meat decreased, given the decline in land for hunting.[65]

4.1.5. Introduction of agricultural techniques and ecological degradation

Food habits were irreparably altered by the introduction of new agricultural techniques which were rapidly assimilated by East African communities for the production of cash crops.[71] These techniques promoted the adoption of higher-yielding food crops such as maize, rice and wheat. These new crops displaced diverse, nutritious, traditional African foods such as millet, sorghum, cowpeas and bambara groundnuts, grown with traditional cultivation techniques.[43, 72] Traditional agricultural techniques included shifting cultivation and intercropping, which evolved to suit local agricultural conditions.[43] These particular cultivation patterns protected the soil, minimized weeds, provided the farm household with a variety of food, and reduced the risk of crop failure, pests and plant diseases.[43]

The systematic decline of indigenous crop varieties resulted in reduced biodiversity and hastened dietary simplification.[71, 73] The shift toward monocultures and reduced dietary diversity has inevitably resulted in a loss of knowledge of ancient agricultural practices.[74] Overall, the introduction of new agricultural methods has benefitted the western powers, but has caused devastation consequences to the people's nutrition and health, as well as to the ecology of East Africa.[75] Numerous plant and animal species are no longer available because habitats have been destroyed, through clearing for commercial agriculture and/or settlement. Traditional foods such as the wild *Dioscorea spp.*, which has historically played a key role in sustaining the population of East Africa during periods of drought and famine, are on the verge of extinction due to the ecological damage incurred.[76]

4.1.6. Displacement of indigenous crops

The onset of cash-crop farming reduced the domestic availability of robust, nutrient-dense, indigenous crops, such as for example sorghum spp., finger millet (*Eleusine coracana*), cowpea (*Vigna unguiculata*), bambara groundnut (*Voandzeia subterranean*), and pigeon pea (*Cajanus cajan* syn. *Cajanus indicus*), all of which are drought tolerant to a considerable extent.[43, 77, 78] The introduction of new dietary staples such as sugar, maize and refined grain flour occurred rapidly, as there was little respect by the colonial powers for the nutritional and cultural benefits of indigenous African foods.[79-81]

4.1.7. Loss of indigenous markets

The indigenous economic system is probably the least understood of all of Africa's social organizations. Although self-sufficiency and subsistence farming were the fundamental foundation of the economic system, Africa did have economies based on agriculture, pastoralism, markets and trade.[6] The marketplace was the heart of pre-colonial indigenous African society. Markets were not only the center of economic activity, but also the center of political, social, judicial and communication activities.[6] In East Africa, studies by Gulliver[82] have revealed that the indigenous markets were extremely important to the *Arusha* people because markets provided them with their "main opportunity for personal contact with the *Maasai* in the conscious efforts to learn and imitate all they could of *Maasai* culture."[82]

Women in Africa have always dominated rural market activity and trade.[83] Local farm produce was almost invariably marketed by women.[83] Today, female traders and some indigenous economic systems still exist. However, when Africa was colonized, the colonialists began to control indigenous economic activities to their advantage.[6] Attempts to reduce or destroy the scale of operations has resulted in the decline of female participation in market activity, sending shock waves through the entire family system, and thewider economic system.[84] Further, the loss of the indigenous marketplace has resulted in reduced access to quality whole foods and reduced indigenous knowledge of deep rooted traditional food habits.

4.1.8. Cultural indoctrination (i.e. "Education")

Indoctrination has negatively impacted on traditional food habits in East Africa since early colonization. This cultural indoctrination has occurred by way of mission schools, boarding schools, and public health programs responsible for educating the youth.[74] These methods of "education" have uniformly reduced knowledge related to the cultivation and preparation of traditional and wild foods.[85] Traditional knowledge has been devalued and waned as the education of children shifted away from the tribal elders, the primary educators in the past, to the imperial powers via the church and school.[68, 86]

"Education" provided by the colonial power encouraged "sophistication" which included a repugnance for traditional foods, and ancient methods of food preparation.[85] For example, the colonialists exhibited contempt for, and prohibited the brewing of native beer rich in vitamin C.[87] Similarly, the use of mineral-rich plant ashes (potashes) as a source of salt used to soften green leaf fibers and make them edible, was condemned.[88] Consequences of this prohibition included the loss of valuable minerals and nutrients from the extract, and the inability to prepare certain traditional dishes.[88] The colonial powers also prohibited the use of many wild plants; this may explain why nutritious wild foods are drastically underutilized today.[70, 89, 90]

4.2. Neocolonial impact on the nutrition transition in East Africa

Colonial influences on Kenya, Tanzania and Uganda abated during the late 1950s to early 1960s when these countries gained independence. The 1960s and 1970s were marked by relative economic growth. However, economic deterioration soon ensued due to a combination of recession in the industrialized countries, rising oil prices and rising interest rates. This meant that numerous developing countries were no longer able to fulfill their governments' and firms' financial obligations.[6, 91] Thus, a need for economic policy reforms was created. The implementation of privatization and cost recovery initiatives (structural adjustment) by the international financial institutions (IFIs), primarily the World Bank and the IMF, from the 1970s through to the present day, has adversely affected the food habits of the East African population and contributed markedly to the deterioration of health status and epidemics of NCDs.[91-93]

One of the core objectives of debt rescheduling in form of the structural adjustment programs and trade liberalization in the 1980s was to "make domestic agriculture more market oriented."[94, 95] Trade policy reforms accelerated in the 1990s as the East African countries liberalized their economies.[4, 6] Invariably, however, economic policy shifts in East Africa over the past 20 to 30 years have resulted in the increased centralization of power to the benefit of the transnational corporations.[96, 97] The corruption underlying the World Bank and IMF programs and the inherent connection with the agenda for global hegemony has been effectively delineated by countless authors.[6, 38, 91, 93, 98, 99]

Adding to the crippling effects of economic reforms is the General Agreement on Tariffs and Trade (GATT).[4] The Uruguay Round of the GATT pledged to improve tariffs, export subsidies, and domestic agricultural support for struggling African countries.[100] However, these measures, once again, led to a restructuring of the national economy which resulted in increased export exploitation by consolidating power amongst a few transnational corporations.[4, 101] These trade policy reforms have enabled greater control by corporations over households through the direct and indirect control of employment opportunities, wages, and daily expenditures, related to subsistence living (i.e. food, clothing, shelter).[101-103]

Delocalization of food production, distribution and marketing has shifted power over food systems from the local economy to a few transnational corporations.[34, 104] In the case of wheat, a handful of companies dominated the world market several decades ago.[105] The world grain market is now determined by one company, Cargill.[106] Four transnational corporations now own approximately 45% of all patents for staple crops such as rice, maize, wheat and potatoes.[107] This rapid centralization of power over the global food system has resulted in swift, commercially driven changes in food habits and tastes[108] and is inherently implicated in the rapid upsurge of NCD epidemics of type 2 diabetes and obesity in East Africa and, indeed, worldwide.

Overall, trade policy reforms have had a devastating impact on traditional food habits throughout Kenya, Tanzania and Uganda. Control exerted over the people of East Africa during the colonial period has not abated since colonial times. Only the technique of control has shifted from the use of overt force to the use of covert force through controlling the political-economic system. Covert control by way of trade policy reforms has fundamentally destroyed the indigenous food habits of East Africa, and has extended the effects of colonial rule *via neocolonial* modus operandi.

4.2.1. Suppression of domestic self-sustainability

The monopolization of agriculture serves the transnational corporations (the ruling power elite) and the western consumer.[3, 32] Indigenous Africans who would normally rely on arable land for food, continue to be displaced or are employed by the corporations for low wages.[73, 75] Instead of growing basic food crops for local people, African farmers are being encouraged to focus on 'high value' agricultural products such as for example fresh flowers and exotic fruits for export.[66]

4.2.2. Dependence on low-quality staple food imports

With all hopes of self-sustainability obliterated by trade policy reforms, East Africans have been increasingly forced to depend on low-quality food imports. In 2001, Africa accounted for 18% of world food imports, up 10% from 1985.[109] According to the FAO,[100] food imports has increased in East Africa due to a decline in agricultural and rural investment.[6]

Wheat currently dominates world food trade,[110] and wheat exports currently come from just five countries: the United States, Canada, Australia, Argentina and France.[110] Increased reliance on imported wheat has been documented in East Africa.[110] Wheat imported into Kenya, Tanzania and Uganda has increased markedly since the early 1990s (Figure 4).[111] Wheat is nutritionally inferior to indigenous East African alternatives, including various millets and sorghum.[20] Moreover, wheat is not a drought resistant crop, unlike millet varieties.[43]

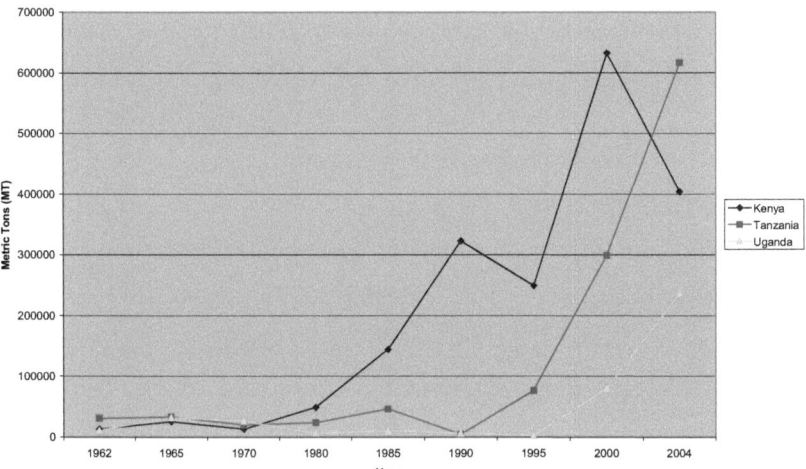

Figure 4. Trends of wheat importation in Kenya, Tanzania and Uganda (1962-2004).[111]

Increased reliance on imported wheat has been attributed to domestic food insecurity, which can in turn be attributed to the trade policy reforms.[32, 112] Import of insidious hydrogenated fats has also increased in East Africa, particularly since the early 1990s. High intake of trans-fatty acids (all types of isomers) increases serum LDL-cholesterol and decrease HDL-cholesterol concentrations,[113] which is associated with an increased risk of CVD.[114, 115] Figure 5 presents data for hydrogenated fat imports from 1994 to 2004 in Uganda.[111] Increased availability and affordability of these particular staple foods versus traditional alternatives has contributed markedly to the dietary simplification and NCDs epidemic in East Africa.[11, 12, 116]

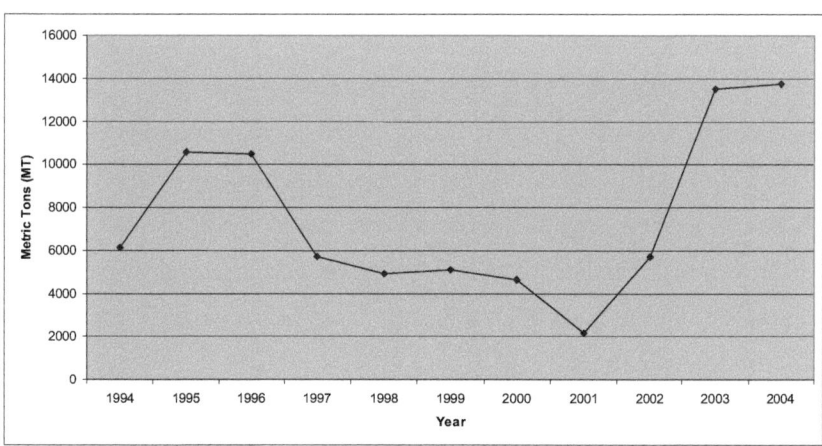

Figure 5. Main imports of hydrogenated oils into Uganda (1994-2004).[111]

4.2.3. Continued displacement of indigenous crops

Recent macroeconomic trade policy reforms have further displaced indigenous crops.[100] In Kenya and Uganda, dietary patterns have shifted from the use of indigenous foods such as millets, sorghums, roots and tubers to greater consumption of vegetable oils[117] (Figure 6-7). Marked reductions in the availability of roots in Kenya (Figure 6) and the reduced availability of millets, sorghum and pulses in Kenya (Figure 6) and Uganda (Figure 7) have been documented.[117] By contrast, wheat, rice, and cheap vegetable fats availability has increased substantially in both Kenya and Uganda during the 1990s.[110] In Tanzania, sorghum and millet availability levels have been generally maintained over the past 60 years (Figure 8).[117] However, a decline in the availability of indigenous starchy roots has been observed since the early 1970s, while the availability of rice and maize have increased (Figure 8).[117]

The governments of many developing countries employ food security measures to ensure that adequate food supplies are available and to keep consumer prices within reasonable limits. However, the implemented measures discriminate in favor of the dominant staple foods (i.e. wheat, maize, and rice) and as such, have shifted food consumption patterns toward nutritionally-inferior crops.[118] Nyoro and Nguyo,[119] have evaluated shifts in food consumption, food production and food purchase patterns induced by the liberalization of maize markets in Kenya, and have revealed that the majority of households are entirely dependent on the market supply of primary staple foods.

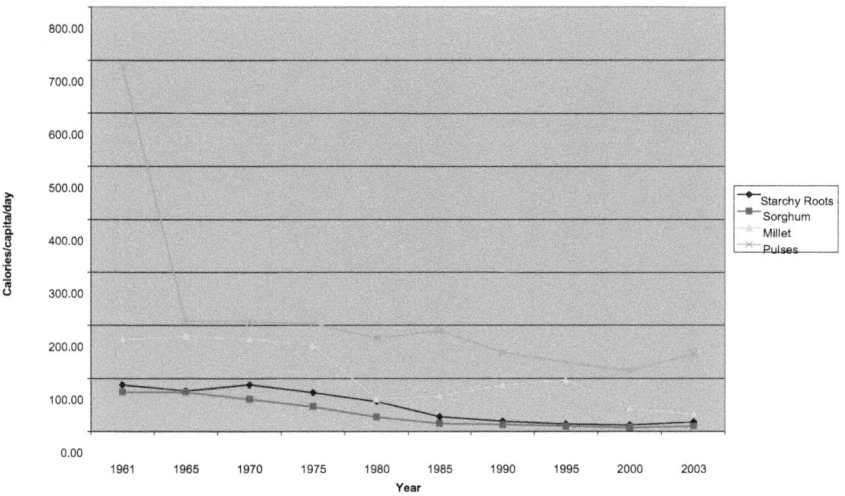

Figure 6. Traditional crops available for consumption in Kenya (1961-2003).[117]

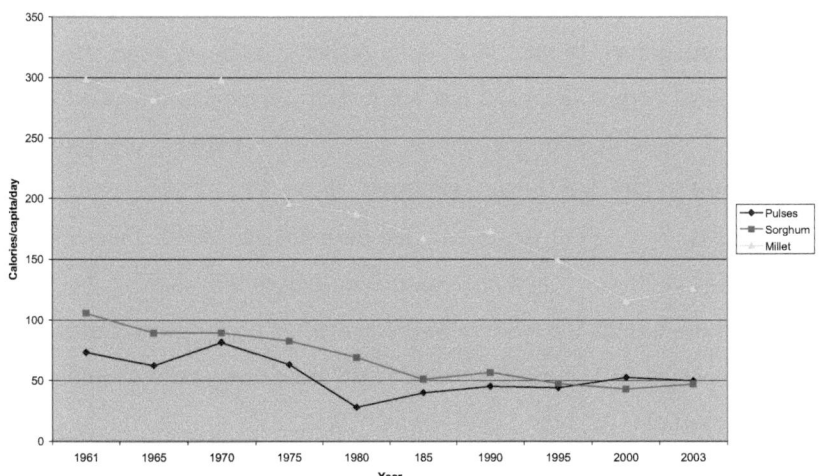

Figure 7. Traditional crops available for consumption in Uganda (1961-2003).[117]

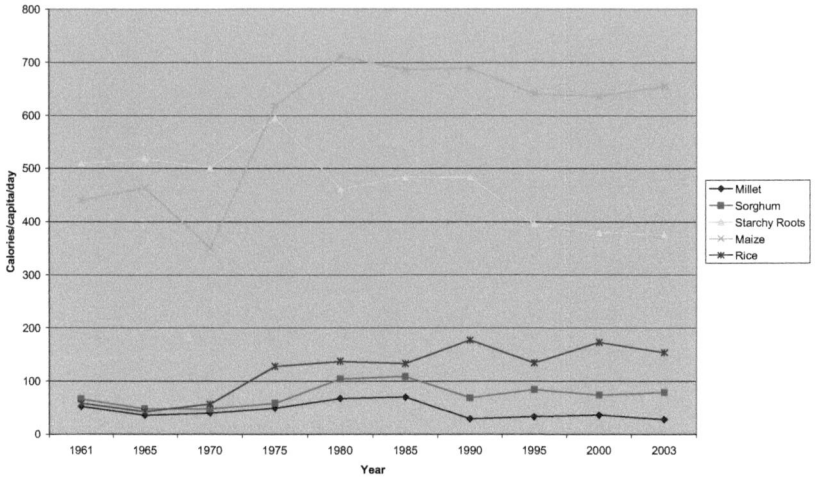

Figure 8. Food crops available for consumption in Tanzania (1961-2003).[117]

4.2.4. Increased availability of processed food products

International trade has consolidated food systems for the transnational corporations and has resulted in the increased availability of processed foods in poorer countries.[72, 97, 120, 121] Further, foreign direct investment (FDI) has been liberalized under GATT and other trade agreements.[122] FDI represents the largest source of financing for developing countries[123] and play an integral key role in shaping the global food culture through the worldwide dissemination of highly processed food products which have been implicated in the *nutrition transition* and global epidemics of obesity, type 2 diabetes and CVD.[121]

In 2001, twelve transnational food product manufacturers ranked among the top 100 list of foreign asset holders worldwide, double the number in 1990.[124, 125] From 1990 to 2002, the combined foreign assets of these companies increased from approximately US$34 billion to US$258 billion.[124, 125] During that same period, the foreign sales of these companies increased from approximately US$89 billion to US$234 billion.[124, 125] Transnational corporations fundamentally drive the integration of world markets and, as such, affect the indigenous and traditional food habits of developing countries.[33] The displacement of indigenous and traditional foods is inevitable given an economic structure which favors these corporations.[126]

The processed foods commonly consumed in East Africa today include cereals, deep fried meat and fish, fried chips, eggs, and many processed sugary and starchy products.[127, 128] Sugary drinks, cakes and doughnuts are amongst the cheapest snacks and are preferred over the traditional roast banana or cassava.[129, 130] Ice cream and ready to eat food items like chocolates, candies and sugary milk products are available for the more affluent.[127] In addition, traditional alcoholic drinks, made from partly germinated cereal flours, honey or fruits, were replaced by industrial beer, wine and hard liquors.[127]

4.2.5. The rise of 'super'markets

Shoprite©, the largest supermarket retailer in Africa has become a deeply entrenched aspect of East African culture.[131, 132] The transformation of food retail, in Africa, including the development and dissemination of supermarkets, first occurred in South Africa, followed by Kenya during the mid 1990s.[133] As they receive substantial FDI from these two countries, Tanzania and Uganda are in the early stages of supermarket development today.[131] In 2002, Kenya had approximately 206 supermarkets and 10 *hypermarkets* (which contain ten times the floor space of a supermarket).[134]

Currently, supermarket expansion is rapidly occurring in Kenya, particularly into poorer demographic areas, and in secondary cities and towns.[131] Supermarkets serve as a primary means to access the highly processed food products of transnational corporations.[135] Access to cheap convenience foods by way of supermarkets has reduced the prevalence of local markets and the availability of indigenous and traditional foods.[131, 133]

4.2.6. Consequences of urbanization

Economic pressures have resulted in mass migration from rural areas to urban centers. Urbanization rates in East Africa are high. In Kenya the percentage of the population living in urban areas increased from 10% in 1970 to 30% in 1997.[136] In Tanzania and Uganda the annual urban population growth rate was 7.4% and 5.9%, respectively, from 1980 to 1990 and 7.1% and 3.8%, respectively, from 1990 until 2000.[137] Several consequences related to rapid urbanization have contributed to the *nutrition transition* in East Africa.

Rapid urbanization in East Africa has contributed to the shifting away from traditional low-fat, high fiber, home produced foods to the preference for pre-prepared, packaged and processed *ready to eat* foods.[129, 138, 139] This increased consumption of saturated fats, sugars, salt and preservatives, minimal fibre and foods low in micronutrients has adverse health implications for the urban East African population.[116, 140-142] Recent evidence from the urban center Dar es Salaam, Tanzania, has revealed a relationship between the consumption of a more *westernized* diet and the increasing prevalence of NCDs, including risk factors which define the metabolic syndrome.[116, 143-147]

The urban East African population is confronted with the widespread availability of packaged food products.[32, 126, 128, 148] The shift in preferences away from traditional, indigenous foods and commodities to diets based on *westernized* consumption patterns has been encouraged, particularly in the last five years by the rapid growth of fast-food restaurants and the explosion of western-style supermarket chains which are rapidly supplying the East Africa urban centers and even rural towns.[131, 149, 150] Many people in urban areas are unaware of the tremendous benefits of indigenous African foods in improving the nutritional status of their daily meals.[48, 151] Also, over time, knowledge of traditional food habits is being lost from one generation to the next, amidst the urban setting.[74]

The extent to which *westernized* dietary patterns and food habits are adopted in *rural* towns has not been effectively investigated. However recent evidence suggests that urban meal patterns are indeed infiltrating the rural areas of Tanzania.[128]

4.2.7. Low-quality non-home prepared foods

Infrastructure in urban living areas is generally characterized by smaller living spaces and poorly equipped kitchens or outdoor cooking spaces, which together with decreased access to natural fuel sources and water, leads to the replacement of many good traditional dietary practices with harmful ones.[152, 153] As a consequence urban consumers become more reliant on highly processed non-home-prepared foods.[127, 152, 154]

Street foods and foods from *kiosks* are the major sources of non-home-prepared foods in the poor urban areas in East Africa.[129, 138, 139] The fact that street foods are inexpensive, time saving and convenient are the main purchasing incentives among poorer, urban East Africans.[130, 139, 155] According to recent studies conducted in Nairobi by van't Riet and colleagues,[130, 138] determinants of non-home-prepared food consumption are derived of a combination of 6 factors, including: (1) the absence of school-age children for women, (2) longer traveling distance to work (3) higher employment status (4) regular income (5) larger household size and (6) higher socioeconomic status. The consumption of street foods may be a factor that parallels socio-cultural changes in food habits, including such behaviors as eating alone, and frequent snacking.[138]

An FAO expert panel[156] has argued that street foods provide cheap and nutritious food that benefit the urban poor. The group hypothesized that many low-income urban families would be even worse off without the availability of street foods. Therefore, the FAO supports the legalization and regulation of this informal food sector activity.[157] A study on street foods sold at construction sites in Nairobi revealed that these particular foods are sold as meals, providing 17-38% of the recommended daily energy intake of manual laborers.[158] Unfortunately however, these street foods contain ingredients which hasten the development of NCDs, including hypertension, obesity and diabetes.[159] Often street foods are prepared using the least expensive ingredients, including highly refined grains and hydrogenated oils. Moreover, these foods contain few essential nutrients, and are high in animal fats, salt, and sugar. In addition, problems of hygiene and food safety are bound to arise in the unsanitary conditions of the shanty towns and slums where the poorest groups live.[154]

4.2.8. Disruption of the family unit

Macroeconomic policy reforms in East Africa have disrupted the family unit by placing greater demands on women. Increasingly, women have been forced to enter the urban labor market to improve family survivability. This has included spending longer hours on the job to meet basic needs.[6] Demands on women living in the rural setting have also increased, particularly with the migration of rural men to urban centers.[160]

The absence of women from the family unit and home has been a factor in reducing the amount of traditional foods, which can be time-consuming to prepare when compared with easily-prepared imported grains and high-calorie/low-nutrient fast foods and street foods.[32, 161] Further, Kennedy et al.[141] have asserted that the increased demands on women are a primary factor for the increased demand of pre-packaged bread in urban Kenya.

Since women have entered the economic employment sector, the breast feeding of infants has declined.[162] A study in urban Morogo, Tanzania, revealed unusually shortened breastfeeding periods.[163] Reduced breast feeding periods are associated with poorer nutritional status and increased susceptibility to diseases such as diarrhea and measles among infants and children.[162]

4.2.9. Disparities of socio-economic status

Gray[164] among others, has argued that with economic liberalization, differences in incomes between various segments of society have increased dramatically, causing marked disparities in food access. Dietary choices are dependent on the socioeconomic status of the family,[65] and lower income families have less access to higher-quality whole foods. Meanwhile, transnational corporations such as McDonalds have become notorious for building the most franchises in the poorest urban areas.[34] The cost of traditional staple foods in urban areas is generally higher than the cost of processed foodstuffs.[165] This has generally resulted in the increased consumption of low nutrient *fast foods* and snacks among the urban poor resulting in dietary deficiencies.[138, 139, 166] Several studies have revealed higher consumption of street foods among urban dwellers of lower socioeconomic status in East Africa.[130, 139]

4.2.10. Nutrition-related propaganda (i.e. advertising)

Nutrition-related propaganda has continued post-colonization with the introduction of mass media, which includes the sophisticated advertising of the transnational corporations. Perhaps the most notorious example of such propaganda was the mass marketing and sale of artificial 'milk' powders for infants and children by Nestlé.[30] The use of these 'milk' formulas reduced the extent of breast feeding, and resulted in deaths from intestinal infections, diarrhea, and dehydration due to contaminated water which was added to the milk formulas.[167]

Today, the mass marketing of packaged food products is ubiquitous and the negative effects of such advertising campaigns have been well documented.[168-170] In the developing world, the marketing strategies often deliberately appeal to existing cultural viewpoints or traditions in order to omit this solely to economically benefit the corporation.[120, 171] Clear contradictions and bizarre connections abound in these advertising campaigns. For example, McDonalds uses it's advertising mashine to promote Unicef and Unicef's mission to eradicate malnutrition among children.[172] Further, AIDS awareness campaigns in Kenya are now promoted on Coca-Cola's billboards.[173]

Children in the developing world have become the primary targets of international marketing campaigns promoting energy-dense, nutrient-poor foods.[174] Advertising is now well-recognized as a significant contributor to the *nutrition transition* and the general acceptance of a *globalized* food culture.[171] Brand marketing has been facilitated by the dramatic improvement in food product distribution within, and between, countries. It is rather ironic that these favorable advances in food distribution have benefited the multinational corporations, yet have not been adequately utilized to facilitate the widespread distribution of quality whole foods to nourish the indigenous people of Africa.[32]

4.2.11. Engineering of famines

While climatic variables play a role in triggering famines, famines in the age of globalization are mainly due to political-economic forces.[5, 175] The IMF-World Bank structural adjustment program bears a direct relationship to the process of famine formation. As outlined by Chossudovsky M.,[176] the pertinent factors in the formation of famines include:

1. Destruction of agriculture and self-sufficiency:

Throughout the 1980s, several austerity measures were imposed on African governments and expenditures on rural development were drastically reduced, leading to the collapse of agricultural infrastructure.[176] Under the structural adjustment programs, farmers increasingly abandoned traditional food crops, as the best land was increasingly allocated to the production of cash crops, used to service western markets and external debt payments.[6]

The structural adjustment programs reinforced dependency on imported grains. Grain imports in sub-Saharan Africa increased from 3.72 million tons in 1974 to 8.47 million tons in 1993.[177, 178] The influx of cheap surplus wheat and rice sold led to major shifts in food consumption patterns, and reduced the demand for traditional crops such as millet and sorghum.[176]

Under the World Bank program, water was to become a commodity to be sold on a cost-recovery basis to impoverished farmers. In the semi-arid regions, this commercialization of water and irrigation, led to the collapse of food security, and ultimately, famine.[177]

Currency devaluation imposed by the IMF in the early 1980s has increased the price of agricultural inputs (i.e. fertilizer, fuel and farm equipment) thus placing a greater burden on debt-ridden farmers. The decline of urban purchasing power during the periods of currency devaluation, the collapse of infrastructure, the deregulation of the grain market, and the influx of food aid has led to the impoverishment of sub-Saharan farming communities.[4, 6]

2. *Collapse of the livestock-economy*

Both the nomadic and commercial livestock of sub-Saran Africa are being destroyed by the IMF and World Bank programs.[4] Animal health has been privatized, and veterinarian services are currently based on user fees. In addition, the structural adjustment programs lead to the absence of emergency feed and water during droughts.[4] As a result, subsidized beef and dairy products, imported duty free from Europe, have led to the demise of the livestock economy of Africa.[179]

3. *Destruction of the state*

The structural adjustment programs systematically undermine all categories of economic activity of the state, which is directly related to the process of famine formation for the following reasons:

- Prevention by the IMF to mobilize domestic resources
- Tight targets of governments for budget deficits
- Donors increasingly provide food aid instead of aid in form of capital and equipment. Food aid in sub-Saharan Africa has increased from 910,000 tons in 1974 to 6.64 million tons in 1993.[177, 178]

4.2.12. Eradication of famine foods

Famines have historically been tolerated through the employment of adaptive responses by the rural people in famine-prone areas. Some of these response mechanisms have been developed in accordance with expected climatic fluctuations encountered seasonally. However, when food shortages extend beyond seasonal fluctuations, vulnerable groups can, among other things, change the composition of their diets.[180, 181] In times of severe famine, families commonly reduce their food intake to one meal a day, as for example in the recent famine due to drought in Tanzania, which resulted in cereal prices around 85 percent higher than average and huge numbers of livestock dead or severely emaciated.[182] Families also try to reduce overall consumption during a famine by diluting gruel with weeds, grass, and other wild greens.[180, 183]

When food supplies lessen drastically, households turn to what are known as *famine foods*. For centuries, indigenous East Africans have adapted their food habits to environmental changes based on their endowment of deep knowledge on drought resistant, indigenous African food crops. Robust drought-resistant crops commonly consumed during environmental challenges include the bambara groundnut, cassava, cowpea, pigeon pea, sorghum and pearl millet.[43] *Famine foods* also grow wild, and include wild vegetables, nuts, berries, and parts of trees. The use and consumption of these plants consumed only at times of food stress is an extremely important adaptive survival strategy.[184]

Historically, the collection and consumption of edible wild foods has enabled people to cope better with erratic untimely rains and drought for several consecutive years without facing severe food shortages, and famine.[184] Wild foods also form an essential part over the 'hunger season' that precedes the harvest[185] and have the potential to become valuable staple foods and important alternatives to the usual food crops cultivated by farmers.[184] Today, many useful wild plants and cultivated indigenous African crops have disappeared due to environmental disturbances brought about by the ecological destruction induced by industrial, commercial, and agricultural development (e.g. cash crop economies), climate change, and urbanization.[186]

4.2.13. Consequences of disease epidemics

In addition to NCDs, communicable disease epidemics related to HIV/AIDS, malnutrition, and infectious diseases continue to escalate in East Africa.[187] Together, or in isolation, these epidemics exert a devastating effect on population health, which drastically alters work and earning capacity, and therefore, food access and consumption patterns. The morbidity and mortality associated with AIDS has been shown to affect farm laboring in the rural setting.[188, 189] Qualitative studies in East Africa[190, 191] suggest that labor shortage and/or labor loss on farms contributes to a change in food habits via one or more of the following 5 factors:

1) Decline in the range of crops grown per farm until only one staple crop is cultivated
2) Change in cropping patterns
3) Decline in livestock numbers leading to a reduced intake of animal protein
4) Loss of traditional knowledge of agricultural and farm management skills, leading to a decline in household food production
5) Disease epidemics markedly increase household expenditures (e.g. for medical treatment, transport, funeral, etc.), resulting in reduced quantity and quality of foods.[192-194] This increases the likelihood of malnutrition in households afflicted with AIDS or other diseases.[195]

In summary, numerous factors have underpinned the *nutrition transition* in the countries of East Africa, including Kenya, Uganda and Tanzania, from early colonization to the current neocolonial forces underlying an oppressive, political-economic structure. It is imperative that greater efforts be directed toward exposing these forces, and proposing solutions to the *nutrition transition* in Africa. Without thorough investigation, documentation and widespread dissemination of this information, efforts to improve the NCD epidemics will prove futile, and the vast continent of Africa, and her people, will continue to suffer.

4.3. Résumé

The non-communicable disease (NCDs) epidemics currently occurring in Africa have been directly attributed to the *nutrition transition*, whereby indigenous foods and food habits have been progressively replaced by the *globalized* food culture of the multinational corporations. Although the simplification of the African cuisine may be most apparent today, the *nutrition transition* has actually taken place over the past 400 years. The purpose of the present chapter was to discuss factors which have underpinned the *nutrition transition* in the countries of East Africa, including Kenya, Uganda and Tanzania, from early colonization to the current oppressive, political-economic structure. It is imperative that greater efforts be directed toward exposing these political-economic forces, and proposing solutions to the *nutrition transition* in Africa. Without thorough investigation, documentation and widespread dissemination of the political-economic forces impeding East African food security, efforts to improve the NCD epidemics will prove futile, and the vast continent of Africa, and its people, will continue to suffer from an increasing rate of chronic diseases.

4.4. References

1. Mutwa V. *Indaba, My Children*. Johannesburg: Grove Publisher, 1999.
2. Dugian P, Gann L, Turner V. *Colonialism in Africa 1870-1960* Cambridge University Press, 1972.
3. Cannon G. Why the Bush administration and the global sugar industry are determined to demolish the 2004 WHO global strategy on diet, physical activity and health. *Public Health Nutr.* 2004; 7(3):369 80.
4. Chossudovsky M. *The Globalization of Poverty and the New World Order* Quebec, Canada: Global Outlook, 2003.
5. Davis M. *Late Victorian Holocausts*: El Nino Famines and the Making of the Third World. London: Verso, 2001.
6. Ayittey G. *Africa Unchanged*: The Blueprint for Africa's Future. New York: Palgrave Macmillan, 2005.
7. World Health Organization. The World Health Report: Today's Challenges. Geneva: World Health Organization, 2003: 1-22.
8. Wild S, Roglic G, Green A, Sicree R, King H. Global prevalence of diabetes: Estimates for 2000 and projections for 2030. *Diabetes Care.* 2004; 27:1047-1053.
9. Popkin B. The nutrition transition and prevention of diet related disease in Asia and the Pacific. *Food Nutr Bull.* 2001; 22:S1-58.
10. Popkin B. An overview of the nutrition transition and its health implications: The Bellagio meeting. *Public Health Nutr.* 2002; 5:93-103.
11. Popkin B. Global nutrition dynamics: The world is shifting rapidly toward a diet linked with noncommunicable diseases. *Am J Clin Nutr.* 2006; 84(2):289-98.

12. Bourne LT, Lambert EV, Steyn K. Where does the black population of South Africa stand on the nutrition transition? *Public Health Nutr.* 2002; 5(1A):157-162.
13. Kitange H, Machibya H, Black J, et al. The outlook for survivors of childhood in sub-Saharan Africa: Adult mortality in Tanzania. *BMJ.* 1996; 312:216-220.
14. Setel P, Unwin N, Alberti K, Hemed Y. Cause-specific adult mortality: Evidence from community-based surveillance - selected sites, Tanzania, 1992-1998. MMWR Morb Mortal Wkly Rep. 2000; 9(19):416-419.
15. Unwin N, Setel P, Rashid S, et al. Noncommunicable diseases in sub-Saharan Africa: Where do they feature in the health research agenda? *Bull World Health Organ* 2001; 79(10):947-953.
16. Popkin B. The nutrition transition and its health implications in lower income countries. *Public Health Nutr.* 1998; 1:5-21.
17. Popkin B. Dynamics of the nutrition transition and its implications for the developing world. *Forum Nutr.* 2003; 56:262-4.
18. Voster H, Bourne L, Venter C, Oosthuizen W. Contribution of nutrition to the health transition in developing countries: A framework for research and intervention. *Nutr Rev.* 1999; 57(11):341-349.
19. World Health Organization. Chronic disease: A vital investment. Geneva: World Heath Organization, 2005.
20. Raschke V, Oltersdorf U, Elmadfa I, Wahlqvist M, Cheema B, Kouris-Blazos A. Content of a novel online collection of traditional East African food habits (1930s - 1960s): Data collected by the *Max-Planck-Nutrition Research Unit*, Bumbuli, Tanzania. *Asia Pac J Clin Nutr.* 2007; 16(1):140-151.

21. Poulter N, Khaw K, Hopwood B, et al. Blood pressure and its correlates in an African tribe in urban and rural environments. *J Epidemiol Community Health* 1984; 38(3):181-5.
22. Poulter N, Khaw K, Hopwood B, et al. Blood pressure and associated factors in a rural Kenyan community. *J Hypertens*. 1984; 6:810-813.
23. Poulter N, Khaw K, Hopwood B, et al. The Kenyan Luo migration study: observations on the initiation of a rise in blood pressure. *BMJ* 1990; 300(6730):967-72.
24. Pauletto P, Puato M, Caroli M, et al. Blood pressure and atherogenic lipoprotein profiles of fish-diet and vegetarian villagers in Tanzania: The Lugalawa study. *Lancet.* 1996; 348(9030):784-8.
25. Pavan L, Casiglia E, Braga L, et al. Effects of a traditional lifestyle on the cardiovascular risk profile: The Amondava population of the Brazilian Amazon: Comparison with matched African, Italian and Polish populations. *J Hypertens*. 1999; 17(6):749-56.
26. Yamori Y, Miura A, Taira K. Implications from and for food cultures for cardiovascular diseases: Japanese food, particularly Okinawan diets. *Asia Pac J Clin Nutr.* 2001; 10(2):144-145.
27. Cockerham W, Yamori Y. Okinawa: An exception to the social gradient of life expectancy in Japan. *Asia Pac J Clin Nutr.* 2001; 10(2):154-158.
28. Trichopoulou A, Critselis E. Mediterranean diet and longevity. *Eur J Cancer Prev.* 2004; 13(5):453-456.
29. Serra-Majem L, Roman B, Estruch R. Scientific evidence of interventions using the Mediterranean diet: A systematic review. *Nutr Rev*. 2006; 64(Suppl 1):27-47.
30. MacLennan R, Zhang A. Cuisine: The concept and its health and

nutrition implications - global. *Asia Pac J Clin Nutr* . 2004; 13(2):131-135.

31. Liu X, Li Y. Epidemiological and nutritional research on prevention of cardiovascular disease in China. Br J Nutr. 2000; 84(Suppl 2):199-203.

32. Lang T. Diet, health and globalization: Five key questions. *Proc Nutr Soc.* 1999; 58(2):335-43.

33. Hawkes C. Uneven dietary development: Linking the policies and processes of globalization with the nutrition transition, obesity and diet-related chronic diseases. *Global Health.* 2006; 2(4):1-18.

34. Schlosser E. *Fast Food Nation*: The Dark Side of the All-American Meal. New York: Houghton-Mifflin, 2001.

35. Eaton S, Konner M. Paleolithic nutrition. *N Engl J Med.* 1985; 312(5):283-90.

36. Robson J. Changing food habits in developing countries. *Ecol Food Nutr.* 1976; 4:251-256.

37. Oniang'o RK. Food habits in Kenya: The effects of change and attendant methodological problems. *Appetite* 1999; 32:93-96.

38. Icke D. *And the Truth Shall set you Free*: The 21st century edition. London: Bridge of Love, 2004.

39. Action AH. *The Geopolitics of Hunger, 2000-2001* Boulder: Lynne Rienner Publishers, 2001.

40. George S. *How the Other Half Dies*: The Real Reasons for World Hunger. Harmondsworth: Penguin, 1976.

41. Maundu P, Imbumi M. East Africa. In: Katz S, Weaver W, eds. *Encyclopedia of Food and Culture.* New York: Thomson and Gale, 2003: 27-34.

42. Maundu P, Ngugi G, Kabuye C. *Traditional Food Plants of Kenya.*

Nairobi: Kenrik, 1999.

43. Gura S. A note on traditional food plants in East Africa: Their value for nutrition and agriculture. *Food Nutr.* 1986; 12(1):18-22.

44. Coupland R. *East Africa and its Invaders.* 2nd edition. London, Great Britain: Oxford University Press, 1956.

45. Tansey G, Worsley T. *The Food System.* London: Earthscan, 1995.

46. Smith B. *The Emergency of Agriculture.* New York: Scientific American Library, 1995.

47. Smith I. *Foods of West Africa*: Their Origin and Use. Ottawa, Canada, 1998

48. O'Ktingati A, Temu R, Kessy J, Mosha T. Indigenous African food crops and useful plants: Their preparation for food, and home gardens in Tanzania. In: Baidu-Forso J, ed. Africa's natural resources conservation and management surveys. Summary Proceedings of the UNU/INRA Regional Workshop. Accra, Ghana: The United Nations University and the Institute for Natural Resources in Africa (UNU/INRA), 1998: 1-3.

49. Ogutu Z. Responding to population pressure in the rural Kenya. *GeoJournal.* 1993; 30(4):409-19.

50. Caldwell C. *The Social Repercussions of Colonial Rule* Demographic Aspects. In: Boahen A, ed. Africa under Colonial Domination 1880-1935. Paris: UNESCO History, 1985: 483.

51. Anderson D. Depression, dust bowl, demography, and drought: The colonial state and soil conversation in East Africa during the 1930s. *Afr Aff.* 1984; 83(332):321-343.

52. Oltersdorf U. Comparison of nutrient intakes in East Africa. The Human Biology of Environmental Change. 1971, Blantyre, Malawi: 51-59.

53. Taylor D. Changing food habits in Kikuyuland. *Can J Afr Stud.* 1970; 4(3):333-349.
54. Bohdal M, Gibbs N, Simmons W. Nutrition surveys and campaigns against malnutrition in Kenya. Report to the Ministry of Health of Kenya. Geneva, Switzerland: WHO/FAO/UNICEF, 1969: 1-171.
55. Courcy-Ireland M, Hosking H, Loewenthal L. An investigation into health and agriculture in Teso, Uganda. In: Committee AS, ed. Nutrition Report. Teso: Agricultural Survey Committee, 1937: 1-28.
56. Jelliffe DB, Benett F, Stroud C, et al. Field survey of the health of Bachiga children in the Kayonza disrtict of Kigesi, Uganda. *Amer J Trop Med Hyg.* 1961; 10:435.
57. Jelliffe D, Jelliffe E, Benett F, White R. Ecology of childhood disease in the Karamojong in Uganda. *Arch Environ Health.* 1964; 9:25-36.
58. Keller R. *Studie zur Ernährung bei zwei Stämmen in Nord-Tanganyika.* Köln, West Germany: Westdeutscher Verlag, 1965.
59. Kreysler J, Schlage C. *The Nutrition Situation in the Pangani Basin.* In: Kraut H, Cremer HD, Ifo Institut für Wirtschaftsforschung e.V. M, eds.Investigations into health and nutrition in East Africa. München: Weltforum Verlag, München, 1969: 85-178.
60. Laurie W, Trant H. A health survey in Kwimba District, Tanganyika. East African Medical Survey. Monograph, East African High Commission. 1954; No. 3.
61. Lema N. Tribal customs in infant feeding II-Among the Chagga. *East Afr Med J.* 1963; 40:370-375.
62. Robson JRK. The district team approach to malnutrition-Maposeni Nutrition Sheme. African Child Health. *J Trop Pediat.* 1962; 8:60.
63. Tanner R. A preliminary enquiry into Sukuma diet in the Lake

Province, Tanganyika territory. *East Afr Med J.* 1956; 33(8):305-324.

64. Winter-Nelson A. International trade and Africa. Illinois: University of Illinois Centre for African Studies, 2004: 1-3.
65. Read M. Native standards of living and African culture change. *Africa.* 1938; 11(Suppl. 3):1-64.
66. Chopra M, Darnton-Hill I. Responding to the crisis in sub-Saharan Africa: the role of nutrition. *Public Health Nutr.* 2006; 9(5):544-50.
67. Turshen M. The impact of colonialism on health and health services in Tanzania. *Int J Health Serv.* 1977; 7(1):7-35.
68. Tabuti J, Dhillion S, Lye K. The status of wild food plants in Bulamogi. County, Uganda. *Int J Food Sci Nutr.* 2004; 55(6):485-498.
69. Food and Agriculture Organization of the United Nations (FAO). *Traditional Food Plants*: A Resource Book for Promoting the Exploitation and Consumption of Food Plants in Arid, Semi-arid and Sub-humid Lands of Eastern Africa. Food and Nutrition Paper 4-2. 1988.
70. Okigbo B. Broadening the food base in Africa: The potential of traditional food plants. *Food Nutr.* 1986; 12(1):4-17.
71. Welch R, Graham R. A new paradigm for world agriculture: Meeting human needs productive, sustainable and nutritious. *Field Crop Res.* 1999; 60:1-10.
72. Shiva V. *Stolen Harvest* London: Zed Books, 2000.
73. Hewitt de Alcantara C. *Modernizing Mexican Agriculture*: Socioeconomic Implications of Technological Change, 1940-1970. Geneva: United Nations Res. Inst. Soc. Dev., 1976.
74. Kuhnlein H, Receveur O. Dietary change and traditional food

systems of indigenous peoples. *Annu Rev Nutr* . 1996; 16:417-42.
75. Tudge C. *So Shall We Reap*. London: Penguin Books, 2003.
76. Asfaw Z, Tadesse M. Prospects for sustainable use and development of wild food plants in Ethiopia. *Econ Bot*. 2001; 55(1):47-62.
77. Harlan J. *Genetic Resources in Africa.* In: Janick J, Simon J, eds. New crops. New York: Wiley, 1993: 65-65.
78. Abegaz B, Demissew S. Indigenous African food crops and useful plants: Their preparation for food, and home gardens in Ethiopia, Kenya, Tanzania and Uganda, with special emphasis on medicinal plants and issues associated with their management. In: Baidu-Forson J, ed. Africa's natural resources conservation and management surveys, Summary proceedings of the UNU/INRA regional workshop. Accra, Ghana: The United Nations University and the Institute for Natural Resources in Africa (UNU/INRA), 1998: 1-4.
79. Allen K. The monotonous diet of the African. *East Afr Med J*. 1955; 32:95-97.
80. Burgess H. Protein-calorie malnutrition in Uganda, II-Busoga District, III-Bukedi District, IV-Bugisu District, V-Ankole district. *East Afr Med J*. 1962;39.
81. Latham M. Malnutrition in East Africa. *J Trop Med Hyg*. 1964; 67:90.
82. Gulliver P. *The Evolution of Arusha Trade*. In: Bohannan, Dalton, eds. Markets in Africa. Evanston: North-Western University Press, 1962.
83. Hodder B. *The Yoruba Rural Markets* In: Bohannan, Dalton, eds. Markets in Africa. Evanston: North-Western University Press, 1962.
84. Skinner E. *Trade and Markets among the Mossi People*. In:

Bohannan, Dalton, eds. Markets in Africa. Evanston: North-Western University Press, 1962.

85. Culwick G. Nutrition in East Africa. *Africa.* 1944; 14(7):401-410.

86. Cattell M. Knowledge and social change in Samia, Western *Kenya. J Cross Cult Gerontol.* 1989; 4(3):225-244.

87. Farnsworth-Anderson. The diet of the African Soldier. *East Afr Agric For J* 1943; 20(7):207-213.

88. Platt B. Tables of representative values of foods commonly used in tropical countries. Medical Research Council, Special Report Series No. 302. London, 1962.

89. Plotkin M. *The Outlook for New Agricultural and Industrial Products from the Tropics.* In: Wilson E, ed. Biodiversity. Washington, DC: National Academy Press, 1988: 106-116.

90. National Research Council. *Lost Crops of Africa.* Vol. 1: Grains. Washington, DC: National Academy Press, 1996.

91. Labonte R, Schrecker T, Sanders D, Meeus W, Cushon J, Torgerson R. *Fatal Indifference*: The G8, Africa and Global Health. Lansdowne and Ottawa: The University of Cape Town Press and the International Development Research Centre, 2004.

92. Breman A, Shelton C. Structural adjustment and health: A literature review of the debate, its role-players and presented empirical evidence. Paper No. WG6:6: Commission on Macroeconomics and Health, 2001.

93. Chossudovsky M. *The Globalization of Poverty*: Impacts of IMF and World Bank Reforms. London: Zed Books, 1997.

94. Milward B. *What is Structural Adjustment?* London and New York: Routledge, 2000.

95. FAO. *The state of Agricultural Commodity Markets* Rome: Food

and Agriculture Organization of the United Nations, 2004: 1-52.
96. Pelto G, Pelto P. Diet and delocalization: Dietary changes since 1750. *J Interdis Hist.* 1983; 14(2):507-28.
97. Lang T. *The Public Health Impact of Globalization of Food Trade.* In: Shetty P, McPherson K, eds. Diet, Nutrition and Chronic Disease: Lessons from Contrasting Worlds. London: J. Wiley & Son, 1997: 173-190.
98. Pilger J. *The New Rulers of the World.* London: Verso, 2003.
99. Klein N. *No Logo*: No Space, no Choice, no Jobs. New York: Picador, 2002.
100. FAO. *The State of Food and Agriculture 2005.* Agricultural Trade and Poverty: Can Trade Work for the Poor? Rome: Food and Agriculture Organization of the United Nations, 2005: 81-93.
101. Lugalla J. The impact of structural adjustment policies on women's and children's health in Tanzania. *Rev Afr Polit Econ.* 1995; 22(63):43-53.
102. ODI. Adjusting to recession: Will the poor recover? ODI Briefing Paper. London: Overseas Development Institute, 1986.
103. Pinstrup-Andersen P. Macroeconomic adjustment policies and human nutrition: available evidence and research needs. *Food Nutr Bull.* 1987; 9(1):1-10.
104. Lang T, Heasman M. *The Global Battle for Mouths, Minds and Markets* London: Earthscan, 2004.
105. Morgan D. *Merchants of Grain.* London: Weidenfeld and Nicolson, 1979.
106. Kneen B. *Cargill and its Transnational Strategies.* Invisible giants. London: Pluto Press, 1995.
107. Action Aid. Going against the Grain, 2003. Available at:

http://www.actionaid.org.uk/_content/documents/gatg_2462004_152 4.pdf#search=%22action%20aid%20going%20against%20the%20gr ain%202003%20pdf%22 (Accessed September 2006).

108. Stitt S, Jepson M, Paulson-Box E, Prisk E. *Schooling for Capitalism: Cooking and the National Curriculum.* In: Koehler B, Feichtinger E, Barlosius E, Dowler E, eds. Poverty and Food in Welfare Societies. Berlin: Sigma, 1997: 363-374.

109. Food and Agriculture Organization of the United Nations (FAO). The State of Food Insecurity in the World. Rome: FAO, 2003: 1-40.

110. FAO. *Financing Normal Levels of Commercial Imports of Basic Foodstuffs*. Commodity Policy and Projections Service, Commodities and Trade Division. Rome: Food and Agriculture Organization of the United Nations, 2003: 1-135.

111. Food and Agriculture Organization of the United Nations (FAO). Key statistics of food and agriculture external trade. Rome: FAO; Available at: http://www.fao.org/statistics/toptrade/trade.asp (Data extracted September 2006).

112. Haan N, Farmer G, Wheeler R. Chronic vulnerability to food security in Kenya 2001. A WFP pilot study for improving vulnerability analysis: World Food Programme (WFP), 2001: 36.

113. Katan M. Trans-fatty acids and plasma lipoproteins. *Nutr Rev.* 2000; 58(6):188–191.

114. Aro A. *Epidemiological Studies of Trans-Fatty Acids and Cardiovascular Disease.* Dundee: The Oily Press, 1998.

115. Oomen C, Ocké M, Feskens E, van Erp-Baart M, Kok F, Kromhout D. Association between trans-fatty acid intake and 10-year risk of coronary heart disease in the Zutphen Elderly Study: A prospective population-based study. *Lancet.* 2001;357(9258):746-751.

116. Njelekela M, Ikeda K, Mtabaji J, Yamori Y. Dietary habits, plasma polyunsaturated fatty acids and selected coronary disease risk factors in Tanzania. *East Afr Med J.* 2005; 82(11):572-8.

117. Food and Agriculture Organization of the United Nations (FAO). Food balance sheets Rome: FAO; Available at: http://www.faostat.fao.org (Data extracted September 2006).

118. FAO. Cereals and other starch-based staples: are consumption patterns changing? Joint Meeting of the Intern-governmental Group on Grains (30th Session) and the Intergovernmental Group on Rice (41st Session). Rome: Food and Agriculture Organization of the United Nations, Economic and Social Department, 2004: 1-17.

119. Nyoro J, Nguyo W. Market liberalization and household nutrition in Kenya: Agricultural Policy, Resource Access and Human Nutrition, 1999.

120. Hawkes C. Marketing activities of global soft drink and fast food companies in emerging markets: A review. Globalization, diets and noncommunicable diseases. Geneva: World Health Organization, 2002.

121. Hawkes C. The role of foreign direct investment in the nutrition transition. *Public Health Nutr.* 2005; 8(4):357-365.

122. Reardon T, Swinnen J. Agrifood sector liberalization and the rise of supermarkets in former state-controlled economies: Comparison with other developing countries: Comparison with other developing countries. Development Policy Review 2004; 22:515-23.

123. Mody A. Is FDI integrating the world economy? *World Econ.* 2004; 27(8):1195.

124. United Nations Conference on Trade and Development (UNCTAD). *World Investment Report.* FDI Policies for Development: National

and International Perspectives. Geneva: UNCTAD, 2003: 322.

125. Organization for Economic Cooperation and Development (OECD). Foreign direct investment, development and corporate responsibility. Paris: OECD, 2000.

126. Cannon G. Nutrition: The new world disorder. *Asia Pac J Clin Nutr.* 2002; 11 (Suppl):498-509.

127. Maletnlema T. A Tanzanian perspective on the nutrition transition and its implication for health. *Public Health Nutr.* 2002; 5(1A):163-168.

128. Mazengo M, Simell O, Lukmanji Z, Shirima R, Karvetti R. Food consumption in rural and urban Tanzania. *Acta Trop.* 1997; 68(3):313-26.

129. Mwangi A, den Hartog A, Foeken D, van't Riet H, Mwadime R, van Staveren W. The ecology of street foods in Nairobi. *Ecol Food Nutr.* 2001; 40(5):497-53.

130. van Riet H, den Hartog A, Mwangi A, Mwadime R, Foeken D, van Staveren W. The role of street foods in the dietary pattern of two low-income groups in Nairobi. *Eur J Clin Nutr.* 2001; 55(7):562-70.

131. Weatherspoon D, Reardon T. The rise of supermarkets in Africa: Implications for agrifood systems and the rural poor. *Dev Policy Rev.* 2003; 21(3):333-355.

132. Kaiza D. Steers, Nandos fast-food chains enter Dar, Kampala. The East African: Nation Group, 1999.

133. Reardon T, Timmer P, Barrett C, Berdegue J. The rise of supermarkets in Africa, Asia and Latin America. *Amer J Agr Econ.* 2003; 85(5):1140-1146.

134. ACNielsen. Sub-Saharan Universe: Kenya. Retail store data. South Africa: ACNielsen, 2002.

135. Neven D, Reardon T. The rise of Kenyan supermarkets and the evolution of their horticulture product procurement systems. *Dev Policy Rev*. 2004; 22(6):669-699.
136. World Bank. *World Development Indicator 1999*. Washington, DC: The World Bank, 1999.
137. FAO. FAOSTAT data. Rome: Food and Agriculture Organization, 2005. Available at: http://www.faostat.external.fao.org/default.jsp (Accessed August 2006).
138. van Riet H, den Hartog A, Hooftman D, Foeken D, van Staveren W. Determinants of non-home-prepared food consumption in two low-income areas in Nairobi. *Nutrition*. 2003; 19(11-12):1006-12.
139. Nasinyama G. *Study on Street Foods in Kampala, Uganda*. Rome and Kampala: Food and Agriculture Organization and Makerere University, 1992.
140. Njelekela M, Kuga S, Nara Y, et al. Prevalence of obesity and dyslipidemia in middle-aged men and women in Tanzania, Africa: relationship with resting energy expenditure and dietary factors. *J Nutr Sci Vitaminol (Tokyo)*. 2002; 48(5):352-8.
141. Kennedy E, Reardon T. Shift to non-traditional grains in the diets of East and West Africa: Role of women's opportunity cost of time. *Food Policy*. 1994; 1:45-56.
142. Edwards R, Unwin N, Mugusi F, et al. Hypertension prevalence and care in an urban and rural area of Tanzania. *J Hypertens*. 2000; 18(2):145-152.
143. Expert Panel on Detection, Evaluation and Treatment of High Blood Cholesterol in Adults. Third report of the National Cholesterol Education Program (NCEP) expert panel on detection, evaluation, and treatment of high blood cholesterol in adults (Adult Treatment

Panel III). *Circulation* 2002; 106:3143 3421.

144. Njelekela M, Negishi H, Nara Y, et al. Cardiovascular risk factors in Tanzania: A revisit. *Acta Trop.* 2001; 79(3):231-9.

145. Njelekela M, Negishi H, Nara Y, et al. Obesity and lipid profiles in middle aged men and women in Tanzania. *East Afr Med J.* 2002; 79(2):58 64.

146. Njelekela M, Sato T, Nara Y, et al. Nutritional variation and cardiovascular risk factors in Tanzania: Rural-urban difference. *S Afr Med J.* 2003; 93(4):295-9.

147. Mosha T. Prevalence of obesity and chronic energy deficiency (CED) among females in Morogoro district, Tanzania. *Ecol Food Nutr.* 2003; 42(1):37-67.

148. Chopra M, Galbraith S, Darnton-Hill I. A global response to a global problem: The epidemic of overnutrition. *Bull World Health Org.* 2002; 80:952-58.

149. Regmi A, Gehlar M. Consumer preferences and concerns shape global food trade - global food trade. *Food Rev* 2001; 24(3):2-8.

150. FAO. Rise of supermarkets across Africa threatens small farmers. FAO News Room (Online). Rome: Food and Agriculture Organization of the United Nations, 2003. Available at: http://www.fao.org/english/newsroom/news/2003/23060-en.html (Accessed August 2006).

151. Kinabo J, Mnkeni A, Nyaruhucha C, Msuya J, Haug A, Ishengoma J. Feeding frequency and nutrient content of foods commonly consumed in the Iringa and Morogoro regions in Tanzania. *Int J Food Sci Nutr.* 2006; 57(1-2):9-17.

152. van Riet H, den H artog A, van Staveren W. Non-home prepared foods: contribution to energy and nutrient intake of consumers living

in two low-income areas in Nairobi. *Public Health Nutr.* 2002; 5(4):515 22.

153. Chandler D, Wane N. Indigenous gendered spaces: an examination of Kenya. Jenda: Journal of Culture and African Women Studies - Online 2002; 2(1): Available at: http://www.jendajournal.com/vol2.1/chandler-wane.html

154. FAO. Food supply systems in Africa. Agriculture food and nutrition for Africa - A resource book for teachers of agriculture. Rome: Food and Agriculture Organization of the United Nations, 1997: 1-32.

155. FAO. Street foods in Nigeria. Comparative study of the socioeconomic characteristics of food vendors and consumers in Ibadan, Lagos and Kaduna. Rome: FAO and Food Basket Foundation International, 1991.

156. FAO. Street foods. Report of an FAO expert consultation Jogjakarta, Indonesia, 5-9 December 1988. FAO Food and Nutrition Paper 46. Rome: FAO, 1990.

157. FAO. Street foods in Nigeria. Comparative study of the socio-economic characteristics of food vendors and consumers in Ibadan, Lagos and Kaduna. Rome: FAO and Food Basket International, 1997.

158. Korir S, Imungi J, Muroki N. Proximate chemical composition of street foods and their energy and protein contribution to the nutrition of manual workers in Nairobi. *Ecol Food Nutr.* 1998; 37:123-133.

159. Simaon S. Promises and challenges of the informal food sector as a strategy for poverty reduction. Report from the Food and Agriculture Organization of the United Nations (FAO) 15 May - 2 June 2006, e-mail conference contributions. Ottawa: Department of Sociology and Anthropology, University of Ottawa, 2006: 22. (Available from: E

mail: ssimaon@uottawa.ca.)

160. Kavishe F, Mushi S. The political economy of nutrition in Tanzania: policy and institutional context. Nutrition-relevant actions in Tanzania. Adelaide: Tanzania Food and Nutrition Centre/UN ACC/SCN, 1993.

161. Fouéré T, Maire B, Delpeuch F, Martin-Prével Y, Tchibindat F, Adoua-Oyila G. Dietary changes in African urban households in response to currency devaluation: Foreseeable risks for health and nutrition. *Public Health Nutr.* 2000; 3(3):293-301.

162. Kavishe F, Mushi S. Food security. Nutrition-relevant actions in Tanzania. Adelaide: Tanzania Food and Nutrition Centre/UN ACC/SCN, 1993.

163. Karegero M. Effects of selected breastfeeding practices on child nutritional status. Paper presented at the Annual Scientific Meeting of the Tanzania Public Health Association. 1989.

164. Gray J. *False Dawn*: The Delusions of Global Capitalism. London: Granta, 1998.

165. Ruel M, Haddad L, Garrett J. Some urban facts of life: Implications for Research and Policy. *World Dev.* 1999; 27(11):1917-38.

166. Kavishe F, Mushi S. Analysis of the nutrition situation and trends in Tanzania. Nutrition-relevant actions in Tanzania. Adelaide: Tanzania Food and Nutrition Centre/UN ACC/SCN, 1993.

167. Akre J. Infant feeding: the physiological basis. *Bull World Health Org.* 1989; 68 (Suppl I):1 -108.

168. Hastings G, Stead M, McDermott L, et al. Review of research on the effects of food promotion to children. Final report prepared for the Food Standards Agency. Glasgow: Food Standards Agency Online, 2003. Available at:

http://www.foodstandards.gov.uk/multimedia/pdfs/foodpromotiontochildren1.pdf#search=%22Review%20of%20research%20on%20the%20effects%20of%20food%20promotion%20to%20children%22 (Accessed August 2006).

169. Hawkes C. Marketing food to children: The global regulatory environment. Geneva: World Health Organization, 2004: 1-75. Available at: http://www.whqlibdoc.who.int/publications/2004/9241591579.pdf#search=%22Marketing%20food%20to%20children%20WHO%22

170. Dalmeny K, Hanna E, Lobstein T. Broadcasting Bad Health. Why food marketing to children needs to be controlled. London: The International Association of Consumer Food Organizations, 2003. Available at: http://whqlibdoc.who.int/publications/2004/9241591579.pdf#search=%22Marketing%20food%20to%20children%20WHO%22

171. Lang T. Obesity: a growing issue for European policy? *J Eur Soc Policy*. 2005; 15(4):301-327.

172. Dyer O. Unicef comes under attack for Big Mac funding deal. *BMJ* 2002; 325:923.

173. McKat B. Coca-Cola to tap its marketing muscle to help fight AIDS epidemic in Africa: Wall Street Journal - online, 2001: available at: http://www.aegis.com/news/wsj/2001/WJ010609.html.

174. James P. Marabou 2005: nutrition and human development. *Nutr Rev.* 2006; 64(S 1-11):72-9.

175. Sen A. *Development as Freedom*: Oxford University Press, 2001.

176. Chossudovsky M. Somalia: The real cause of famine. The globalization of poverty and the new world order, Canada: *Global Outlook* 2003: 95 -102.

177. World Bank. *World Development Report 1992*: Development and the Environment. New York: Oxford University Press, 1992.
178. Food and Agriculture Organization of the United Nations (FAO). Food supply situation and crop prospects in sub-Saharan Africa. Special Report Series No. 1. Rome: FAO, 1993:10.
179. World Bank. Sub-Saharan Africa: From crisis to sustainable growth. Washington, DC: World Bank, 1989.
180. Longhurst R. Famines, food, and nutrition: issues and opportunities for policy and research. *Food Nutr Bull*. 1987; 9(1):1-11.
181. BBC. Famine threat across the Horn. UK: BBC News World Edition-Online, 2000. Available at: http://www.news.bbc.co.uk/2/hi/africa/696803.stm (Accessed September 2006)
182. United Nations World Food Programme. Tanzania drought: urgent food aid for half a million. News - Press Release Online. Dar Es Salaam: United Nations World Food Programme, 2006. Available at: http://www.wfp.org/English/?ModuleID=137&Key=2093 (Accessed September 2006)
183. Biellik R, Henderson P. Mortality, nutritional status and dietary conditions in a food deficit region: North Teso District, Uganda. *Ecol Food Nutr*. 1981; 11(163-70).
184. Maundu P, Njiro E, Chweya J, Imungi J, Seme E. Kenya. In: Chweya J, Eyzaguirre P, eds. *The Biodiversity of Traditional Vegetables*. Rome: International Plant Genetics Resource Institute (IPGRI), 1999: 51-82.
185. Guinand Y, Lemessa D. Wild-food plants in Ethiopia. Reflections on the role of 'wild-foods' and 'famine-foods' at a time of drought: UN-Emergencies Unit for Ethiopia (UNDP-EUE), United Nations

Development Programme, 2000: 1-20.
186. Bell J. The hidden harvest. Seeding, The Quarterly Newsletter of Genetic Resources Action International, 1995: 8.
187. Johns T. Dietary diversity, global change, and human health. Proceedings of the symposium "Managing Biodiversity in Agricultural Ecosystems". 2001, Montreal, Canada: 1-11.
188. World Health Organization. Diet, nutrition and the prevention of chronic diseases. Report of a Joint WHO/FAO Expert Consultation. In: WHO, ed. Report of a Joint WHO/FAO Expert Consultation. Geneva, 2003: 1-148.
189. WHO. The world health report 2004 - changing history. Rome: World Health Organization, 2004: 1-156.
190. Barnett A, Rugalema G. HIV/AIDS. 2020 Focus No. 05 - Brief 03: International Food Policy Research Institute, 2001. Available at: http://www.ifpri.org/2020/focus/focus05/focus05_03.asp (Accessed October 2006)
191. Barnett T. The effects of HIV/AIDS on farming systems and rural livelihoods in Eastern Africa. Summary analysis. Rome: FAO, 1994.
192. Topouzis D, Hemrich G. The socio-economic impact of HIV/AIDS on rural families in Uganda. UNDP Discussion Paper No. 6. New York: UNDP, 1996.
193. Rugalema G. AIDS and rural livelihoods: A case study of smallholder households of Bukoba District, Tanzania. PhD Thesis: The Hague, 1998.
194. Barnett T, Haslwimmer M. The effects of HIV/AIDS on farming systems in Eastern Africa. Rome: FAO, 1995.
195. Tumushabe J, Bantebya G, Ssebuliba R. The effects of HIV/AIDS on agricultural production and rural livelihood systems in Eastern

Africa. FAO Project RAF/92/TO/A/08/12. Uganda: FAO, 1993.

196. de Waal A, Tumushabe J. HIV/AIDS and food security in Africa. A report for DFID (Department for International Development): Southern African Regional Poverty Network - Online, 2003: 1-22. Available at: http://www.sarpn.org.za/documents/d0000235/P227_AIDS_Food_Security.pdf (Accessed August 2006)

5. IS THE NON-COMMUNICABLE DISEASE EPIDEMIC IN EAST AFRICA BEING CAUSED BY A LOSS OF TRADITIONAL FOOD HABITS?

5.1. Objectives for the systematic review

At the beginning of the 19th century, non-communicable diseases (NCDs) were virtually non-existent amongst the indigenous populations of East Africa.[1,2] Today, NCDs have reached epidemic levels.[3-8] Hypertension is highly prevalent,[3,4] and co-exists with cardiovascular disease (CVD)[7,8] and type 2 diabetes in both urban and rural regions of East Africa.[6] It is likely that within the next 20 years, sub-Saharan Africa will experience a three-fold increase in deaths due to CVD and a near three-fold increase in the incidence of type 2 diabetes.[9] The World Health Organization (WHO) has recently revealed that NCDs currently account for nearly 80% of deaths in developing countries.[10] This statistic is notable as NCD epidemics are reported increasingly across the developed world.

The recent, rapid rise of NCDs throughout East Africa has in large part been attributed to a *nutrition transition,* whereby traditional foods and food habits have been progressively replaced by the *globalized* (or *westernized*) food culture of the transnational corporations.[11-13] The first major shift in food habits in East Africa reportedly occurred during the 1920s[1,14] when commercial salt was added to meals, progressively replacing native ash rich in potassium salt. This seemingly minor shift in the diet has had significant repercussions. For example, the addition of commercial salt has been associated with a greater prevalence of hypertension in East Africa.[1,14,15] Further, the Cardiovascular Disease and Alimentary Comparison (CARDIAC) study[16] has suggested that even the most rudimentary dietary shifts such as increased intake of commercial salts, and reduced intake of potassium and polyunsaturated fatty acids (PUFA), can contribute markedly to the NCD epidemics in East Africa.[15, 17]

Chapter 4 reported on the colonial and *neocolonial* forces responsible for the eradication of traditional foods and food habits throughout East Africa. Throughout recent history, the *globalized food culture* has systematically replaced the traditional foods and ancient, indigenous knowledge of the practical, cultural, and medicinal uses of these foods. The *nutrition transition,* largely implicated in the genesis of NCDs,[18, 19] is indeed not limited to the addition of commercialized salts to the East African cuisine.[20] Overall, the *globalized* food culture has resulted in dietary simplification, which includes a diet high in animal (saturated) fats, trans-fatty acids, cheap vegetable oils, sugary foods and beverages, highly-processed foods, and nutritionally inferior staple crops (i.e. wheat, rice and maize).[21, 22] Moreover, the *globalized* food culture has led to reduced dietary intake of fibre, and whole foods, fruits, and vegetables.[21, 22] The medical and socioeconomic consequences of the NCD epidemics resulting from this *nutrition transition* will be disastrous for the East African countries, where impoverished health care systems are already overburdened.[23]

Nutrition is coming to the forefront as a major modifiable determinant of NCDs, with scientific evidence increasingly supporting the view that long-term improvement of food habits can have a profound, positive effect on health status throughout life. The public health approach of primary prevention, which includes a locally diverse and traditional diet, is considered to be the most cost-effective, affordable and sustainable course of action to reduce the upward trend of NCDs.[21]

We have recently provided evidence regarding the myriad health benefits of traditional East African food habits (See Chapter 2).[22] Further, according to a survey conducted at the 18th International Congress on Nutrition (ICN) in Durban, South Africa, 2005, 84% of experts in the nutritional sciences (n=92) believed that traditional African foods and food habits were superior to the *globalized* food culture currently underpinning the *nutrition transition*.[24] The purpose of this investigation was to determine if adherence to a traditional East African diet is associated with better markers of health status, including a lower NCDs risk factor profile, versus adherence to a non-traditional diet.

5.2. Methods for the systematic review

A systematic, critical review rather than a meta-analytic approach has been undertaken as the heterogeneity of outcome measures do not lend themselves to meta-analytic methods.

5.2.1 Criteria for considering studies

Study designs

Quantitative cross-sectional investigations involving at least one cohort consuming a traditional African diet were included in the present review. "Traditional" can be defined as indigenous or introduced foods and food habits which due, to a very long use, have been assimilated as part of the culture of the community. Each study had to include at least one group from East Africa, including Kenya, Tanzania, and/or Uganda. Intervention trials and case reports were excluded.

Subjects

Subjects were randomly selected adult (≥18 years) men and women. Subjects were all free of diagnosis of priority diseases of the main NCD cluster, including: cardiovascular disease (i.e. coronary heart disease and cerebrovascular accident), diabetes mellitus, cancer, and chronic respiratory diseases. Studies which did not meet these criteria were excluded.

Outcome measures

Studies which evaluated both dietary intake *and* health status indices were included. Dietary intake may have been quantified for example using dietary recall assessments, weighted food records, and/or dietary intake questionnaires (e.g. food frequency questionnaires). Health status may have been quantified for example using anthropometrics (e.g. BMI, skinfold measurements), haematological assessments (e.g. blood lipid profiles, insulin resistance, fasting blood glucose) and/or clinical assessments (e.g. blood pressure). Studies not meeting these criteria were excluded.

Statistical analyses

Studies performing statistical analyses between groups and/or correlations and regressions between any dietary intake and health status variables of interest were included. Studies not performing these statistical analyses were excluded.

5.2.2. Search protocol

A literature review was conducted from May to October 2006 from the years 1959 to 2006, limited to the English language, using computerized databases including Ovid Medline, Web of Science, Pubmed and Google Scholar. The search combined keywords related to NCDs (i.e. cardiovascular disease, hypertension, type 2 diabetes mellitus), NCD risk factors (i.e. obesity, overweight, atherogenic dyslipidemia, insulin resistance, glucose intolerance, proinflammatory state, prothrombotic state), diet (i.e. rural/urban diet, food consumption, dietary factors, food frequency, cross sectional dietary studies, dietary intake, food habits, and dietary habits) and East Africa (i.e. Kenya, Uganda and Tanzania). Articles retrieved were examined for further relevant references.

5.3. Results of the systematic review
5.3.1. Studies excluded and included

The systematic review process is presented in Figure 9. Of 37 potentially relevant articles, 29 were excluded for the following reasons. Twenty four did not evaluate dietary intake,[3, 4, 6, 12, 13, 15, 17, 25-41] two investigations did not evaluate health status[42, 43], two investigations were incomplete in their study results (i.e. dietary factors)[11, 44] and one study applied an intervention.[45]

Eight cross-sectional studies were included in the present review (Figure 1).[46-53] All eight studies evaluated dietary intake ad health status.[46-53]

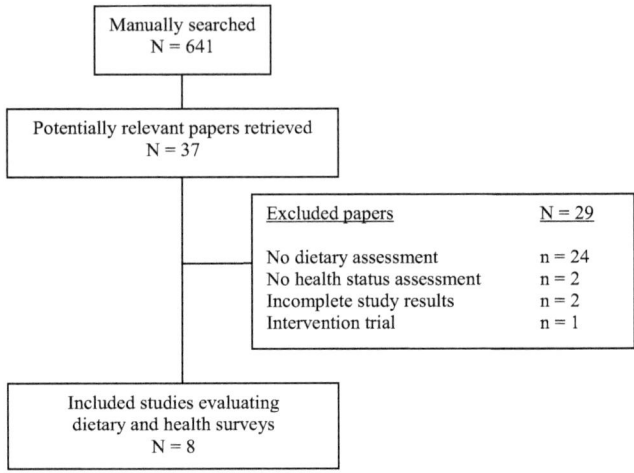

Figure 9. Flow and characteristics of publications included/excluded for review.

5.3.2. Overview of subjects

Sample sizes

Five-thousand four-hundred and ninety-eight (n = 5498) subjects were randomly selected and enrolled in the eight studies reviewed.[46-53] Sample sizes ranged from 105 to 1263 enrolled subjects. Only one investigation enrolled fewer than 400 subjects (n=105),[49] while four studies enrolled 445 to 608 subjects (Table 10).[46-48, 50] and three studies enrolled greater than 970 subjects.[51-53]

Gender

Two studies did not provide a gender breakdown.[51, 52] Of the remaining studies, all except one (which included 976 women only)[53] enrolled both men and women. In total, 999 men and 2126 women (ratio of 32:68) were enrolled in the seven investigations which provided a gender breakdown.[46-53]

Age

One study did not provide information regarding the age of their sample.[52] In five studies, age was expressed separately for women and men. Mean age ranged from 50.7 to 52.2 years for women and from 50.4 to 52.0 years for men, respectively, in three studies expressing age as mean ± standard deviation,[46-48] and from 37.3 to 52.3 years for women and from 41.6 to 53.0 years for men, respectively, in two studies expressing age as mean ± standard error of estimate.[49, 50] One study presented an age range where the youngest and eldest subjects enrolled were 27.0 and 41.0 years, respectively.[53] In another trial, age was presented as mean ± standard deviation for each Tanzanian cohort enrolled: the Lupingu village was 38.0 ± 14.2 years and the Madilu village was 46.6 ± 13.1 years.[51]

5.3.3. Overview and categorization of cohorts

Urban Dar es Salaam, Rural Handeni and Pastoral Monduli, Tanzania[46-49]

Four studies by Njelekela *et al.*[46-49] involved cohorts from urban Dar es Salaam, rural Handeni and the pastoral Monduli communities of Tanzania. The cohort from urban Dar es Salaam primarily consisted of middle-income civil servants, petty traders and middle-class residents with moderate incomes. The subjects of rural Handeni were primarily employed as subsistence agriculturalists, and had a lower income compared to the urbanites. This cohort lived in a traditional African village and consumed primarily a traditional vegetarian diet. The semi-nomadic Maasai pastoralists of Monduli tended large herds of cattle and were not involved in any agricultural activities. The Maasai were lean and tall, had an active lifestyle, and consumed a traditional diet.[46-49]

Categorization

    Traditional East African diet: Handeni and Monduli

    Non-traditional diet: Dar es Salaam

Lupingu and Madilu communities, Tanzania[50, 51]

Two studies involved the Lupingu and Madilu communities of Tanzania, whose foods sources were derived from fishing and farming, respectively.[50, 51] Both communities maintained a traditional lifestyle without piped water, electricity, telephone, or television, and spoke the same Kishwahili dialect.[50, 51] In the Lupingo village, fish accounted for approximately one-quarter of the total caloric intake, whereas residents of the Madilu village primarily consumed a vegetarian diet. Both cohorts adhered to their traditional diets. Cigarettes and canned food and drinks were rarely if ever available for consumption.

Categorization

    Traditional East African diet: Lupingo and Madilu

    Non-traditional diet: None

Lugarawa district, Tanzania, and the Lugbara, Uganda[52]

Pavan *et al.*[52] investigated two cohorts from East Africa, including the Lugarawa district of Tanzania and the *Lugbara* community of Uganda, and compared these cohorts to residents of the Amazonian region of Brazil (i.e. Ouro Preto do Oeste council area, Rodonia) and the northern Italy (Mirano, Venice, and Castelfranco, Veneto) (Table 10).[52] The diet of the

African cohort was based on fresh-water fish, vegetables and fruits, with low consumption of saturated fats and salt (4 grams per day).[52] The Brazilian cohort had just begun the transition from a rural to an urban lifestyle, with an associated shift in dietary patterns toward a more *westernized* diet. Incomes and level of education of this cohort was low. The general lifestyle of the Italian cohort was typical of an urbanized and industrialized area.

Categorization

Traditional East African diet: Lugarawa, Tanzania and *Lugbara*, Uganda

Non-traditional diet: Italian and Brazilian

Investigations of women in the Morogoro district of Tanzania[53]

Mosha *et al.*[53] stratified their study population into four cohorts based on self-reported occupation: (1) farmers, (2) business women, (3) housewomen/housewives, and (4) civil servants. Further, businesswomen were classified as medium or high level depending on their capital investment.

Categorization

Traditional East African diet: Farmers and house-women/house-wives

Non-traditional diet: Civil servants and business women (medium and high)

5.3.4. Methodologies

Dietary intake assessments

Dietary intake was evaluated by a 7-day food frequency questionnaire,[46-52] a 1-day food frequency questionnaire[53] and the evaluation of dietary intake from standardized food tables.[50-52] Twenty-four hour (24hr) urine collection was used to evaluate sodium chloride (NaCl) as an index of salt intake.[46, 48]

Health status assessments

Studies evaluated systolic and diastolic blood pressure,[46, 48, 49, 51, 52] anthropometric measurements such as body weight,[46-49, 51-53] height,[46-49, 51-53] hip and waist circumference,[47, 49] body fat (i.e. sum of skinfold thickness)[50] and body mass index (BMI, kg/m^2).[46-53] Twenty-four hour resting energy expenditure (REE) was evaluated in kcal/min/kg by indirect calorimetry.[46, 47] Twenty-four hour (24hr) urine collection was used to evaluate the intake of sodium (Na),[46, 48] potassium (K),[46, 48] magnesium (Mg),[46, 48] calcium (Ca)[46] and urea nitrogen.[48] Haematological assessments included the evaluation of total cholesterol (TC),[46-49, 51, 52] triglycerides (TG),[47, 51] high-density lipoprotein cholesterol (HDLC),[46-49, 51] low-density lipoprotein cholesterol (LDLC),[47, 49] total protein,[46] lipoprotein (a) (Lp(a)),[51] glycosylated haemoglobin (HB$_{A1c}$),[47-49] plasma insulin,[50] blood glucose [50-52] plasma leptin,[50] and fatty acids (FA) in plasma phospholipids such as saturated fatty acids (SAFA) (i.e. lauric acid,[49] myristic acid,[49] palmitic acid,[49, 51] stearic acid)[49, 51]), monounsaturated fatty acids (MUFA) (i.e. oleic acid,[49, 51] and mead acid[51]), polyunsaturated fatty acids (PUFA) (i.e. linoleic acid,[49, 51] -linoleic acid,[49, 51] dihomo-gamma-linolenic acid (DGLA),[51] arachidonic acid (AA),[49, 51] docosapentaenoic acid (DPA),[51] total ω-6 FA,[49, 50] eicosapentaenoic acid (EPA),[49, 51] docosahexaenoic (DHA),[49, 51] total ω-3 FA).[49-51]

Statistical analyses

Statistical analyses were performed as follows:
- Five trials performed statistical comparisons between groups on outcomes related to dietary intake[46-50]
- Six trials performed statistical comparisons between groups on outcomes related to health status[47-52]
- Seven trials investigated relationships between dietary and health status outcomes[46-50, 52, 53]

5.3.5. Outcomes

Outcomes related to dietary intake and health status in each group investigated are presented in Table 10.

- <u>*Investigations of Dar es Salaam, Handeni and Monduli, Tanzania*[46-49]</u>

Dietary intake

Njelekela *et al.* reported that the consumption of coconut milk, meat, and fish (primarily deep-fried) was significantly higher (among both men and women) in urban Dar es Salaam as compared to their counterparts in Handeni and Monduli (Table 10).[46,49] Consumption of whole milk was reported to be significantly higher among the pastoral Maasai population of Monduli than their counterparts in Dar es Salaam and Handeni.[47-49] Intake of green vegetables was significantly higher in rural Handeni compared to their urban (Dar es Salaam) and pastoral (Monduli) counterparts (Table

10).[48, 49] Twenty-four hour urine revealed higher sodium chloride excretion and sodium/potassium ratio among men and women in urban Dar es Salaam versus counterparts in Handeni and Monduli (p<0.05) (Table 10).[48]

Health status

Hypertension (systolic and diastolic) was more prevalent among urban Dar es Salaam men[48, 49] and women[48] versus their rural and pastoral counterparts (Table 10).[48] Women from Handeni were determined to have significantly higher systolic and diastolic blood pressure than Monduli women, however neither group was hypertensive (128.3 ± 23.2 mmHg and 111.5 ± 19.1 mmHg, respectively)[48]

Dar es Salaam men and women had significantly higher BMI[46-49] and prevalence of obesity[46-48] as compared to their rural and pastoral counterparts. Prevalence of central obesity was significantly higher in urban Dar es Salaam men (21.4%) as compared to Handeni (1.4%) and Monduli men (0%) (Table 10; p<0.001).[47] Prevalence of central obesity was high among women (ranging from 42.9 to 57.9%) and was not significantly different between the three groups investigated.[47]

In one study HB_{A1c} percentage was significantly higher among men from Monduli as compared to the men of Handeni (p<0.05).[48] Another trial revealed that the HB_{A1c} was significantly higher among men of Dar es Salaam and pastoral Monduli as compared to the men of Handeni (p<0.05).[49] HB_{A1c} among women showed the same pattern (i.e. significantly higher in Dar es Salaam (p<0.05) and Monduli (p<0.001) as compared to Handeni) (Table 10).[48]

Prevalence of hypercholesterolaemia was highest among the men and women of Monduli as compared to the other two cohorts.[46] Overall, the prevalence of hypercholesterolaemia was lowest among the Handeni men and women.[46, 48] The men[47-49] and women[47, 48] of Dar es Salaam and Monduli had significantly higher TC[47-49] and LDLC concentrations versus their counterparts in Handeni.[47, 49] HDLC levels were significantly lower in the men and women of rural Handeni as compared to the other two groups (Table 10).[47, 48] Dyslipidemia (i.e. (TC-HDLC)/HDLC) was significantly higher among Handeni women (p<0.05) and Monduli men (p<0.05) as compared to their counterparts.[47] Overall, TC (p<0.001)[47], LDLC (p<0.001)[47] and the prevalence of hypercholesterolaemia (p<0.001)[46] were significantly higher in women than in men across all groups investigated.

Myristic acid was significantly lower in Handeni men and women as compared to Dar es Salaam (p<0.05).[49] A significant difference in the plasma concentrations of PUFA were also observed: AA (C20:4 ω-6) was significantly higher in Handeni as compared to Dar es Salaam and Monduli.[49] DHA (C22:6 ω-3) was significantly lower among the Monduli as compared to Dar es Salaam and Handeni.[49] Total ω-3 FA concentration was lower among the Monduli than their urban and rural counterparts.[49]

In two surveys,[46, 47] the REE adjusted for body weight was significantly higher among men in Monduli and Handeni than in Dar es Salaam (Table 10; p<0.001). In women, REE adjusted for body weight was significantly higher in Monduli than in Dar es Salaam (Table 10; p=0.004).[46, 47]

- *Investigations of Lupingu and Madilu communities, Tanzania*[50, 51]

Dietary intake

Total caloric intake was similar between the Lupingu and Madilu (Table 10).[50, 51] Salt intake was also similar between groups.[51] In the Lupingo group, 23% of total energy intake was derived from non-deep-fried fish (300-600 g/day; 3-4 fish meals/day), whereas fish constituted approximately 6% of the Madilu diet (Table 10). Amongst the Madilu, most of the energy was derived from complex carbohydrates (82%) such as maize and rice, while protein and fats constituted 11% and 7% of the diet respectively. The Madilu did not eat meat. The Lupingu derived 70% of their dietary intake from complex carbohydrates, 18% from protein and 12% from fat. Consumption of local alcoholic beverages (obtained by the fermentation of cereals; 4-6% alcohol content) was significantly greater amongst the Madilu, as compared to the Lupingu (Table 10).[50, 51]

Health status

The Lupingu cohort (fish diet) had significantly lower systolic and diastolic blood pressure and significantly lower prevalence of borderline ($p<0.0001$) and definite hypertension ($p<0.0001$) compared to Madilu (vegetarian diet).[51] Overweight and obesity were not highly prevalent (0.2 to 4.5% prevalence) in either cohort.[51] Assessment of subcutaneous body fat among these particular two groups revealed no significant difference between Lupingo and Madilu men (Table 10), whereas Lupingo women had significantly higher subcutaneous body fat as compared to Madilu women ($p=0.018$).[50] However, Lupingo women had significantly lower fasting insulin levels ($p=0.011$) as compared with Madilu women.[50]

Overall, the Lupingo cohort (men and women combined) had significantly lower plasma lipid concentrations (i.e. TC, TG and HDLC; $p<0.0001$) versus the Madilu cohort.[51] The Lupingo (fish diet) group also had a significantly lower prevalence of hypercholesterolaemia ($p<0.0001$) and hypertriglyceridaemia ($p<0.0001$) (Table 10).[51] Lipoprotein a (Lp(a)) concentrations were 37% lower in the Lupingo group as compared to the Madilu ($p<0.0001$). The percentage of ω-3 PUFA was significantly higher among the Lupingo (p<0.001). The percentage of ω-6 FA was lower in the Lupingo than the Madilu (p<0.001), while the percentage of AA was significantly higher (Table 10; p<0.005). The ω-3 to ω-6 ratio, as well as the ratios of EPA and DHA to AA, were nearly four times higher in the Lupingo versus the Madilu (Table 10; p<0.001).[51]

- *Investigations of the Lugarawa district of Tanzania, the Lugbara community of Uganda, Italy and Brazil*[52]

Dietary intake

Pavan *et al.*[52] compared groups residing in the rural Lugarawa district of Tanzania and the *Lugbara* community of Uganda (combined data) with groups residing in Brazil and Italy. Daily energy intake in the African groups (9360 J) was slightly lower than the Brazilian (9860 J) and Italian groups (10750 J) (no statistical comparison drawn). Consumption of meat, sugar, dairy products and oil was much lower in the African cohort versus the Brazilian and Italian cohorts (Table 10). Salt (NaCl) intake was 4 g per day in Africa, and 10 g per day in Brazil and Italy (Table 10; p<0.0001).[52]

Health status

The African cohort had lower blood pressure (systolic and diastolic) versus the Brazilian and Italian groups. Specifically, systolic blood pressure of the African group was 8% lower than that of the Brazilian group and 11% lower than that of the Italian group (both p<0.0001). Diastolic blood pressure was 13% lower in the Africans versus both other groups (p<0.0001).[52] BMI, cholesterol, glycaemia, and prevalence of overweight women were significantly lower in the African group as compared to the other two cohorts (Table 10).

- *Investigations of women in the Morogoro district of Tanzania[53]*

Dietary intake

The frequency of food intake per person per day investigated among the female groups engaged in different socioeconomic activities in rural and urban areas of the Morogoro district, Tanzania, ranged from 2 to 4 times per day for the obese subjects and 1 to 4 times for the non-obese (BMI between 18.5-24.9) subjects. Average food intake per person was 3 times per day for females in all categories.[53]

Health status

Prevalence of obesity across all female groups was 49%, while 47% were categorized as normal (BMI of 18.5-24.9). Prevalence of obesity was highest among civil servants and lowest among the farmers.[53]

Relationship between dietary intake measures and health status (NCD risk factors)

Significant relationships between dietary intake and health status variables of interest are presented in Table 10.

Dietary intake and blood pressure

Using the pooled data of all men in their studies, Njelekela et al.[46, 48] demonstrated positive associations between sodium-potassium ratio and systolic blood pressure and sodium-potassium ratio and diastolic blood pressure (Table 10). Sodium-potassium ratio was also positively correlated with systolic blood pressure in women (Table 10; $p<0.05$).[46]

Njelekela et al.[48] revealed that frequency of meat, fish (deep-fried), and coconut milk consumption were all positively correlated with both systolic and diastolic blood pressure among women in their investigation (Table 10). This particular study[48] also revealed positive relationships between the frequency of meat consumption and both systolic and diastolic blood pressure among men (Table 10). In another study by Njelekela et al.,[49] frequency of fish intake in both men and women were positively correlated with both systolic and diastolic blood pressure (Table 10).

Pavan et al.[52] revealed positive associations between blood pressure and alcohol consumption in the African, Brazilian, and Italian groups investigated (Table 10). In the Brazilian participants, both systolic and diastolic blood pressure increased with the westernization (i.e. globalization) of dietary habits.[52] Blood pressure was lower in those eating mainly a vegetable-based diet than in those eating a meat-based diet.[52]

Dietary intake and body mass index (BMI)

Njelekela et al.[46, 47, 49] revealed positive associations between BMI and the intake of meat, coconut milk, and fish amongst both genders investigated (Table 10). Multiple regression analysis revealed a positive relationship between BMI and meat,[46, 47] fish, and coconut milk intake among both genders[46] and in men only (Table 10).[47] By contrast, BMI was inversely related to the intake of whole milk in women (Table 10; $p<0.001$),[47] and the intake of green vegetables in both genders investigated (Table 10).[48] In a study by Mosha et al.[53] BMI was positively associated with the frequency of food intake in obese subjects.

Dietary intake and blood lipid profiles

Njelekela et al. have revealed several important relationships between dietary intake measures and indices of health status. In several trials, serum TC was inversely related to frequency of fish consumption in men (Table 10).[47-49] LDLC was also inversely related with the frequency of fish consumption in men.[49] By contrast, serum TC was positively associated with meat intake in men[47, 48] and women[47] (Table 10). Linoleic acid (C 18:2 ω-6) was inversely related to fish consumption among men (Table 10).[49] AA and EPA were positively related to the frequency of fish consumption in women (Table 10).[49] There was a positive relationship between DHA and fish intake in both genders (Table 10).[49] Total ω-3 FA was positively related to frequency of fish consumption among women (Table 10).[49] Ratio of ω-3 to ω-6 FA was positively associated with frequency of fish intake in both genders (Table 10).[49]

Winnicki et al.[50] revealed that leptin levels were positively related to alcohol consumption in the men enrolled in their trial (Table 10; p=0.041). Leptin levels were also independently associated with alcohol consumption in men and women consuming vegetable diets.[50] In multivariate analysis among the fish and vegetarian subgroups, plasma leptin levels were independently associated with type of diet (p<0.001).[50]

In a study by Pavan et al.[52], serum cholesterol levels were positively related to frequency of meat intake. Further, red meat intake was more responsible for hypercholesterolaemia than white meat (p<0.02).[52]

Glycosylated haemoglobin

In a study by Njelekela et al.,[49] HB_{A1c} percentage was positively associated with frequency of fish intake in both men and women (Table 10; p<0.05)

5.4. Discussion

The purpose of this systematic review was to determine if adherence to a traditional East African diet was associated with better markers of health status, including a lower NCDs risk factor profile, versus adherence to a non-traditional diet. The systematic review process resulted in eight investigations which evaluated the dietary intake and health status of cohorts residing in Tanzania and Uganda.[46-53] All eight studies included in the review primarily focused on health status outcome measures, and therefore provide limited evidence regarding the characteristics of the traditional diet consumed by the particular East African cohort studied. Given this limitation, we can only draw minimal conclusions between the importance of the traditional East African diets and their health benefits related to the prevention of NCDs.

Our review provides some evidence for the protective role of a traditional East African diet on the NCD risk factor profile, including hypertension, dyslipidaemia and obesity. Studies reviewed also provide some support for the benefits of a high fish diet on these NCD risk factors, particularly blood lipid profiles.[50-52]

Six surveys involved cohorts adhering to a *westernized* diet and/or experiencing a *nutrition transition*.[46-49, 52, 53] The cohort from urban Dar es Salaam consumed a diet higher in salt and saturated fats, including coconut milk, meat, and deep-fried fish, and lower in vegetables compared to their pastoral and/or rural counterparts.[46-49] These dietary considerations, coupled with lower levels of physical activity, may have contributed to the elevated blood pressure, increased prevalence of obesity and central obesity investigated among urban men versus the comparison groups (Handeni and Monduli).[46-49] These findings are in agreement with other investigations,[42] which have suggested that residents of urban Dar es Salaam have undergone, and continue to undergo, a *nutrition transition* toward a diet involving a higher consumption of fats (saturated and trans-fats), salt, and sucrose and lower intake of dietary fibre.

Our review revealed higher TC, LDLC and FA levels in urban Dar es Salaam and pastoral Monduli as compared to rural Handeni, which was concomitant with a greater consumption of high-fat foods amongst the urban and pastoral groups.[47-49]

Increased prevalence of dyslipidemia has been observed in several trials of urban Tanzania, where socioeconomic status tends to be higher than that of rural regions.[4, 54] A non-vegetarian diet has been implicated, given that individuals of higher socio-economic status tend to consume more animal foods high in saturated fat.[4] Swai *et al.*[4] determined that hypercholesterolaemia, serum cholesterol and BP were higher and more prevalent in the more developed, urban areas of Kilimanjaro as compared to the rural areas, and that this may have been related to a greater consumption of meat and dairy products in the urban centers.

In addition, the high intake of coconut milk, as revealed in the urban Dar es Salaam cohort investigated by Njelekela *et al.*[47-49] may be a further contributor to the higher NCD risk factor prevalence in urbanites. Also, Mazengo *et al.*[42] observed higher intake of coconut milk and SAFA levels in urban Dar es Salaam. The chronic use of saturated fats and oils may have detrimental effects on the health status of the people in Dar es Salaam, considering that SAFA (palmitic C16:0 and stearic C18:0 acids) found in coconut tends to elevate serum TC and LDLC concentrations.[55]

The Maasai of Monduli have long been known for their very low prevalence of cardiovascular disease despite their high consumption of saturated fat.[44] Hypercholesterolaemia and elevated HB_{A1c} percentage among the pastoral Maasai community, as revealed by Njelekela *et al.*[48] may be due to the fact that their dietary habits are undergoing a transition. The elevated risk of NCDs observed may be attributed to a greater intake of dietary fat and/or decreased consumption of traditional foods including fermented milk[56] and plant additives, which contain significant amounts of hypocholesterolaemic saponins and phenolics.[57] These foods have long helped the Maasai to maintain low cholesterol levels.[58] However, despite the elevated NCD risk factors revealed among the Massai in the present review, the REE of Maasai women and men was significantly higher than that of rural (Handeni) and urban (Dar es Salaam) counterparts.

Our review revealed a high prevalence of obesity among women in East Africa.[41, 59] In a trial by Njelekela *et al.*[47] the prevalence of central obesity ranged from 42.9% to 57.9% among the three female groups investigated (Table 10). Further, the women exhibited significantly greater TC,[47] LDLC,[47] and prevalence of hypercholesterolaemia[46] as compared to the men enrolled in these particulalr trials. The survey conducted in the Morogoro district of Tanzania[53] also revealed a high prevalence of obesity (28 to 65%) among all women investigated. Although, the frequency of food intake per day among the *obese* and *normal* (BMI 18.5-24.9 kg/m^2) females was similar, it was observed that the female civil servants and businesswomen had a habit of consuming high fat foods and non-diet carbonated drinks such as Coke and Pepsi, which were commonly sold as snacks in their work places.[53]

These findings are in line with the recent published evidence,[60] suggesting that the high frequency of eating and snacking does not necessarily predispose individuals to be overweight or obese. It is the *qulity* of food, or rather, a lack thereof, which is predisposing obesity. We have recently investigated the causes underlying a *globalized* food culture and the connection to non-communicable disease epidemics (See Chapter 4). High incomes (among civil servants and businesswomen)[53] and easy access to high fat foods and sweetened drinks, could be one of the major factors contributing to the high prevalence of obesity among urban females.[53]

Overall, NCDs risk factors were lower in rural Handeni than in urban Dar es Salaam and pastoral Monduli.[46-49] This finding may be supported by an increased consumption of green vegetables among the rural population. Several investigations draw attention to the process of *westernization* in dietary habits as a trigger factor for NCDs.[11, 27, 61] Shortly after indigenous populations came into contact with the Western culture, their NCD patterns changed dramatically.[11, 12] More recently this has been observed in Chimbu (New Guinea), in subjects from the Cook Islands (Melanesia),[62] in the Tarahumara (Mexico),[63] the Pima Indians (Arizona),[64] and in the Brazilian cohort[52] included in our review, who have just began the transition from a rural to urban lifestyle.

The study conducted in the Lugarawa district of Tanzania and among the *Lugbara* community of Uganda,[52] revealed that a traditional East African diet is related to a favorable NCD risk factor profile, as compared to other global dietary habits, including those of Brazil and Italy. The investigations in the Lupingo and Madilu villages in Tanzania[50, 51] pointed out the beneficial effects of an increased fish consumption on the NCD risk factor profile among the fish diet group of Lupingo village. The favorable levels and low prevalence of NCD risk factors observed among the fish diet group compared to the survey community living on a vegetarian diet, was attributed to the high concentration of ω-3 PUFA, included in fish oil, and their antithrombotic action and modification of immunological processes.[65, 66]

In particular, the degree of fish consumption observed among the fish diet group in Lupingo[50, 51] (300-600g/day), was one of the highest found in population studies, compared to Greenland Eskimos[67] and the Japanese fishermen of Ushibuka,[68, 69] investigated in the Seven Countries study,[69] who consumed an average of 400 g/day of seal, whale or fish and 200g/ day of fish, respectively. The associated benefits of the fish diet investigated in Lupingo[50, 51] included lower BP, favorable lipid profiles and the reduction in plasma leptin concentration.[51] Therefore, these particular findings signify the favorable risk profile of a diet rich in ω-3 PUFA, originally described among Eskimos.

In conclusion, the review revealed that there is only limited data available regarding the relationship between dietary factors and NCD risk factors among East African cohorts adhering to traditional and non-traditional dietary patterns. Considering the potential health benefits of a traditional lifestyle including the practice of indigenous dietary habits, modification of undesirable (*westernized*) dietary habits may be important for the prevention of NCDs in at-risk East African populations. Further work is needed to clearly identify traditional diets among East African population groups and the magnitude and impact of the *nutrition transition* on dietary change and its relation to the increasing prevalence of NCDs in East Africa.

5.5. Résumé

The purpose of this investigation was to determine if adherence to a traditional East African diet was associated with better markers of health status, including a lower non-communicable disease (NCD) risk factor profile, as compared with adherence to a non-traditional diet. A systematic review of computerized databases was performed. Studies which evaluated both dietary intake *and* health status indices were included. The search resulted in eight quantitative cross-sectional investigations involving at least one cohort consuming a traditional African diet. The studies included in the review provide limited information regarding the intake of micro and macronutrients and the composition of meals in the cohorts studied, making the data difficult to interpret. However, the studies reviewed provide some support for the health related benefits of the traditional East African diet versus a non-traditional diet, particularly with regard to NCD risk factors such as hypertension, dyslipidaemia and obesity. The studies reviewed also provide some support for the protective effects of increased fish consumption, particularly blood lipid profiles. Additional research is needed to more thoroughly document traditional diets amongst the East African population, and investigate relationships between dietary intake and health status indices. Such research is needed to identify the magnitude and impact of the *nutrition transition* on food habits and the prevalence of NCD in East Africa.

Table 10: Characteristics of eight dietary assessments and health status surveys on non-communicable disease (NCD) risk factors.

Authors[a,c] (year) country	N	Study groups location (n)	Dietary intake outcomes			Health status outcomes			Relationship between dietary variables and NCD risk factors		
			Variable	Value or prevalence	P value	Variable	Value or prevalence	p	(Study groups) NCD risk factor vs. dietary variable	r	P value
Njelekela et al[46] (2001) Tanzania	446	Women:									
		Urban Dar es Salaam (n=79)	Food frequency intake:						Women:		
			Meat consumption	highest	NR[a]	Obesity[1] (%)	43.6		SBP vs. Na:K ratio	+0.159	<0.05
			Fish consumption	highest	NR[a]	Hypertension[2] (%)	61.7		BMI vs. meat intake	+0.929	<0.0001
			Coconut milk consumption	highest	NR[a]	Hypercholester- olaemia[3] (%)	50.0		BMI vs. fish intake	+0.600	<0.0001
						REE	0.017±0.004		BMI vs. coconut milk	+0.543	<0.0001
		Rural Handeni (n=91)				Obesity[1] (%)	15.4		Men:		
						Hypertension[2] (%)	29.8		SBP vs Na:K ratio	+0.270	<0.0001
						Hypercholester- olaemia[3] (%)	22.5		DBP vs. Na:K ratio	+0.251	<0.0001
						REE	0.019±0.005		BMI vs. meat intake	+0.458	<0.0001
		Pastoral Monduli (n=61)				Obesity[1] (%)	6.8		BMI vs. fish intake	+0.305	<0.0001
						Hypertension[2] (%)	19.1		BMI vs. coconut milk	+0.440	<0.0001
						Hypercholester- olaemia[3] (%)	53.7				
						REE	0.020±0.004				
		Men:	Food frequency intake:								
		Urban Dar es Salaam (n=81)	Meat	highest	NR[a]	Obesity[1] (%)	8.6				
			Fish	highest	NR[a]	Hypertension[2] (%)	63.1				
			Coconut milk	highest	NR[a]	Hypercholester- olaemia[3] (%)	9.5				
						REE	0.019±0.005				
		Rural Handeni (n=93)				Obesity[1] (%)	2.1				
						Hypertension[2] (%)	26.9				
						Hypercholester- olaemia[3] (%)	4.4				
						REE	0.024±0.006				
		Pastoral Monduli (n=41)				Obesity[1] (%)	2.4				
						Hypertension[2] (%)	25.2				
						Hypercholester- olaemia[3] (%)	48.6				
						REE	0.024±0.006				

Table 10 (continue): Characteristics of eight dietary assessments and health status surveys on non-communicable disease (NCD risk factors.

Authors[Ref] (year) country	N	Study groups location (n)	Dietary intake outcomes			Health status outcomes				Relationship between dietary variables and NCD risk factors		
			Variable	Value or prevalence	P value	Variable	Value or Prevalence	p	(Study groups) NCD risk factor vs. dietary variable	r	P value
Njelekela et al[67] (2002) Tanzania	545	Women:									
		Urban Dar es Salaam (n=92)	Food frequency intake:						Women:		
			Green vegetable (d/wk)	4.2±2.3	<0.001[b]	BMI	28.6±6.7	<0.001[a]	BMI vs. meat intake	+0.344	<0.001
			Coconut milk (d/wk)	4.0±2.6	<0.001[a]	Obesity[1] (%)	40.0	<0.001[a]	BMI vs. fish intake	+0.289	<0.001
			Whole milk (ml/day)	69.6±137.3		Central Obesity[1] (%)	51.6		BMI vs. coconut milk consumption	+0.350	<0.001
			Fish (d/wk)	3.1±2.1	<0.001[c]	TC[2]	5.3±1.2	<0.001[c]	BMI vs. whole milk intake	-0.246	<0.001
			Meat (d/wk)	2.5±1.8		TG[4]	3.7±1.9	<0.001[a]	TC vs. meat intake	+0.183	=0.002
						LDLC[5]	3.2±1.0				
						HDLC[5]	1.3±0.4	<0.001[c]	Men:		
						Dyslipidemia[5] (%)	7.8	<0.05[c]			
						REE	0.017±0.004	=0.004[b]	BMI vs. meat intake	+0.263	<0.001
		Rural Handeni (n=107)	Green vegetable (d/wk)	4.8±2.1*	<0.001[b]	BMI	24.3±6.5		BMI vs. fish intake	+0.210	=0.001
			Coconut milk (d/wk)	2.1±2.0*		Obesity[1] (%)	17.0		BMI vs. coconut milk consumption	+0.345	<0.001
			Whole milk (ml/day)	85.9±92.9*		Central Obesity[1] (%)	42.9		TC vs. meat intake	+0.272	<0.001
			Fish (d/wk)	2.1±1.6*		TC[2]	4.4±1.5				
			Meat (d/wk)	1.8±1.8*		TG[4]	2.2±1.0				
						LDLC[5]	3.0±1.4				
						HDLC[5]	1.0±0.4				
						Dyslipidemia[5] (%)	19.0				
						REE	0.019±0.005				
		Pastoral Monduli (n=87)	Green vegetable (d/wk)	1.7±2.4*		BMI	21.0±5.0				
			Coconut milk (d/wk)	0.1±0.8*	<0.001[a]	Obesity[1] (%)	10.7				
			Whole milk (ml/day)	785.1±841.4*	<0.001[a]	Central Obesity[1] (%)	57.9				
			Fish (d/wk)	0.1±0.4*		TC[2]	5.3±1.2	<0.001[c]			
			Meat (d/wk)	2.5±2.2*		TG[4]	1.7±0.8				
						LDLC[5]	3.7±1.1	<0.001[c]			
						HDLC[5]	1.3±0.4	<0.001[c]			
						Dyslipidemia[5] (%)	9.0	<0.05[c]			
						REE	0.020±0.004				

Table 10 (continue): Characteristics of eight dietary assessments and health status surveys on non-communicable disease (NCD)risk factors.

Authors[Ref] (year) country	N	Study groups location (n)	Dietary intake outcomes			Health status outcomes			Relationship between dietary variables and NCD risk factors (Study groups) NCDs risk factor vs. dietary variable	r	P value
			Variable	Value or prevalence	P value	Variable	Value or Prevalence	p			
Njelekela et al[47] (2002) Tanzania	545	**Men:** Food frequency intake:									
		Urban Dar es Salaam (n=89)	Green vegetable (d/wk) Coconut milk (d/wk) Whole milk (ml/day) Fish (d/wk) Meat (d/wk)	3.7±2.3 3.9±2.8 48.3±118.0 2.7±1.8 2.3±1.8	<0.001[a]	BMI Obesity[1] (%) Central Obesity[1] (%) TC[4] TG[4] LDLC[5] HDLC[5] TC-HDLC/HDLC Dyslipidemia[5] (%) REE	25.2±4.3 10.3 21.4 4.8±1.2 3.6±2.5 2.8±1.1 1.2±0.4 3.1±1.5 9.6 0.019±0.005	<0.001[a] <0.05[a] <0.001[a] <0.001[c] <0.001[a] <0.001[c] <0.05[c] <0.001[b] <0.001[a]			
		Rural Handeni (n=111)	Green vegetable (d/wk) Coconut milk (d/wk) Whole milk (ml/day) Fish (d/wk) Meat (d/wk)	5.1±2.2 1.7±2.1 134.9±773.6 2.3±2.0 1.5±1.6	<0.001[a] <0.001[b]	BMI Obesity[1] (%) Central Obesity[1] (%) TC[4] TG[4] LDLC[5] HDLC[5] TC-HDLC/HDLC Dyslipidemia[5] (%) REE	21.3±3.2 2.7 1.4 3.4±0.9 2.1±1.4 2.0±1.0 1.0±0.7 3.0±1.8 7.3 0.024±0.006	<0.05[b]			
		Pastoral Monduli (n=59)	Green vegetable (d/wk) Coconut milk (d/wk) Whole milk (ml/day) Fish (d/wk) Meat (d/wk)	1.1±1.7 - 1218.6±1679.0 - 2.0±2.4	<0.001[a] <0.001[a]	BMI Obesity[1] (%) Central Obesity[1] (%) TC[4] TG[4] LDLC[5] HDLC[5] TC-HDLC/HDLC Dyslipidemia[5] (%) REE	20.7±3.2 1.7 0.0 5.1±1.5 1.9±1.5 3.7±1.3 1.1±0.4 4.8±5.2 22.6 0.024±0.006	<0.001[c] <0.001[a] <0.05[a]			

Table 10 (continue): Characteristics of eight dietary assessments and health status surveys on non-communicable disease (NCD) risk factors.

Authors[Ref.] (year) country	N	Study groups location (n)	Dietary intake outcomes			Health status outcomes			Relationship between dietary variables and NCD risk factors		
			Variable	Value or prevalence	P value	Variable	Value or prevalence	p	(Study groups) NCD risk factor vs. dietary variable	r	P value
Njelekela et al[68] (2003) Tanzania	445	Women:	Food frequency intake:[‡]						Women:		
		Urban Dar es Salaam (n=79)	Green vegetable (d/wk)	4.3±2.3	<0.0001[a]	SBP	141.3±27.5	<0.001[a]	BP vs. meat intake	NR	<0.05
			Coconut milk (d/wk)	4.1±2.6		DBP	81.2±21.4	<0.001[a]	BP vs. fish intake	NR	<0.05
			Whole milk (ml/day)	65.8±1304.8		Hypertension[2] (%)	52.0	<0.0001[a]	BP vs. coconut milk consumption	NR	<0.05
			Fish (d/wk)	3.1±1.9		BMI	29.3±6.7	<0.0001[a]	BP vs. coconut milk consumption	NR	<0.001
			Meat (d/wk)	2.5±1.9	<0.05[c]	Obesity[1] (%)	45.5	<0.0001[a]	BMI vs. green vegetable intake	NR	<0.05
						TC[4]	5.3±1.2				
						Hypercholester-olaemia[3] (%)	40.6				
						HDLC	1.3±0.4		Men:		
						Hb$_{A1C}$[6]	5.2±1.2	<0.05[c]	BP vs. meat intake	NR	<0.05
						Na excretion	highest	<0.05[a]	BP vs. /Na/K intake	NR	<0.05
						Na/K ratio	highest	<0.05[a]	TC vs. meat intake	NR	<0.05
		Rural Handeni (n=91)	Green vegetable (d/wk)	4.9±2.1		SBP	128.3±23.2	<0.0001[b]	TC vs. fish intake	NR	<0.05
			Coconut milk (d/wk)	2.0±1.9		DBP	74.6±12.9	<0.05[b]	BMI vs. green vegetable intake	NR	
			Whole milk (ml/day)	85.6±198.6		Hypertension[2] (%)	34.5				
			Fish (d/wk)	2.1±1.6		BMI	24.2±6.9	<0.001[b]			
			Meat (d/wk)	1.8±1.7		Obesity[1] (%)	17.8				
						TC[4]	4.3±1.3	<0.0001[a]			
						Hypercholester-olaemia[3] (%)	22.5	<0.0001[a]			
						HDLC	1.0±0.4	<0.0001[a]			
						Hb$_{A1C}$[6]	4.8±0.7				
		Pastoral Monduli (n=61)	Green vegetable (d/wk)	1.5±1.3	<0.0001[a]	SBP	111.5±19.1				
			Coconut milk (d/wk)	0.1±0.9		DBP	67.4±14.9				
			Whole milk (ml/day)	804.9±890.2	<0.0001[a]	Hypertension[2] (%)	15.5				
			Fish (d/wk)	0.1±0.3	<0.0001[a]	BMI	20.1±4.2				
			Meat (d/wk)	1.1±1.9		Obesity[1] (%)	5.6				
						TC[4]	5.4±1.2				
						Hypercholester-olaemia[3] (%)	47.6				
						HDLC	1.3±0.5				
						Hb$_{A1C}$[6]	5.6±0.5	<0.001[c]			

Table 10 (continue): Characteristics of eight dietary assessments and health status surveys on non-communicable disease (NCD)risk factors.

Authors[Ref.] (year) country	N	Study groups location (n)	Dietary intake outcomes			Health status outcomes			Relationship between dietary variables and NCD risk factors		
			Variable	Value or prevalence	P value	Variable	Value or prevalence	p	(Study groups) NCD risk factor vs. dietary variable	r	P value
Njelekela et al[48] (2003) Tanzania	445	**Men:**	**Food frequency intake:**								
		Urban Dar es Salaam (n=81)	Green vegetable (d/wk)	3.7±2.3		SBP	140.4±24.8	<0.001[a]			
			Coconut milk (d/wk)	3.9±2.7	<0.0001[a]	DBP	79.7±16.6	<0.001[a]			
			Whole milk (ml/day)	45.6±105.9		Hypertension[2] (%)	53.1	<0.0001[a]			
			Fish (d/wk)	2.7±1.8		BMI	24.9±4.4	<0.0001[a]			
			Meat (d/wk)	2.4±1.8	<0.05[c]	TC[4]	4.9±1.2				
						Hypercholesterolaemia[3] (%)	29.9				
						HDLC[6]	1.3±0.4				
						Hb$_{A1C}$[6]	5.3±1.5				
						Na excretion	higher	<0.05[a]			
						Na/K ratio	higher	<0.05[a]			
		Rural Handeni (n=93)	Green vegetable (d/wk)	4.9±2.2*		SBP	112.2±18.3				
			Coconut milk (d/wk)	1.7±2.1*		DBP	69.1±10.2				
			Whole milk (ml/day)	154.9±845.3*		Hypertension[2] (%)	16.1				
			Fish (d/wk)	2.4±1.9*		BMI	21.1±2.9				
			Meat (d/wk)	1.5±1.5*		TC[4]	3.4±0.9	<0.05[a]			
						Hypercholesterolaemia[3] (%)	4.4				
						HDLC[6]	1.0±0.7	<0.0001[a]			
						Hb$_{A1C}$[6]	4.8±1.1	<0.05[a]			
		Pastoral Monduli (n=40)	Green vegetable (d/wk)	1.1±1.7	<0.0001[a]	SBP	116.8±16.7				
			Coconut milk (d/wk)	-		DBP	68.9±13.0				
			Whole milk (ml/day)	1307.5±1844.9	<0.0001[a]	Hypertension[2] (%)	28.6				
			Fish (d/wk)	-	<0.0001[a]	BMI	20.6±3.4				
			Meat (d/wk)	1.7±2.5		TC[4]	5.3±1.3				
						Hypercholesterolaemia[3] (%)	50.7				
						HDLC[6]	1.1±0.4				
						Hb$_{A1C}$[6]	5.7±0.5	<0.05[c]			

Table 10 (continue): Characteristics of eight dietary assessments and health status surveys on non-communicable disease (NCD) risk factors.

Authors[Ref.] (year) country	N	Study groups location (n)	Dietary intake outcomes			Health status outcomes			Relationship between dietary variables and NCD risk factors		
			Variable	Value or prevalence	P value	Variable	Value or prevalence	p	(Study groups) NCD risk factor vs. dietary variable	r	P value
Njelekela et al[79] (2005) Tanzania	105	**Women:**	**Food frequency intake:**			**All sub-groups:**			**Women:**		
		Urban Dar es Salaam (n=14)	**Dar es Salaam**			**Dar es Salaam**			BMI vs. fish intake	+0.306	<0.0001
			Fish <1d/wk (%)	5.9		Myristic acid (%)	0.7±0.1	<0.05[c]	SBP vs. fish intake	+0.379	<0.0001
			Meat>2d/wk (%)	58.8		AA (%)	5.8±0.4	<0.05[c]	DBP vs. fish intake	+0.320	<0.0001
			Green vegetable>2d/wk (%)	83.3		DHA (%)	1.4±0.2	<0.0001[b]	Hb$_{A1c}$ vs. fish intake	-0.153	<0.05
			Coconut milk>2d/wk (%)	72.2	<0.0001[b]	Total ɷ-3 FA (%)	2.8±0.3	<0.0001[b]	AA vs. fish intake	+0.392	<0.05
			Fruits>2d/wk (%)	61.1	<0.0001[b]	**Handeni**			EPA vs. fish intake	+0.379	<0.05
			Whole milk>400mls/d (%)	5.7		Myristic acid (%)	0.3±0.1		DHA vs. fish intake	+0.493	=0.0010
		Rural Handeni (n=16)	**Handeni**			AA (%)	7.0±0.2		Total ɷ-3 vs. fish intake	+0.423	<0.05
			Fish <1d/wk (%)	13.9		DHA (%)	1.2±0.1		-3/ɷ-6 ratio vs. fish intake	+0.423	<0.05
			Meat>2d/wk (%)	38.9	<0.0001[a]	Total ɷ-3 FA (%)	2.5±0.2	<0.05[b]	EPA+DHA vs. fish intake	+0.491	<0.05
			Green vegetable>2d/wk (%)	94.4	<0.0001[a]	**Monduli**					
			Coconut milk>2d/wk (%)	61.1		Myristic acid (%)	0.5±0.1		**Men:**		
			Fruits>2d/wk (%)	27.8		AA (%)	4.8±0.3	<0.001[c]	BMI vs. fish intake	+0.227	<0.05
			Whole milk>400mls/d (%)	2.8		DHA (%)	0.5±0.1	<0.001[c]	SBP vs. fish intake	+0.194	<0.05
		Pastoral Monduli (n=11)	**Monduli**			Total ɷ-3 FA (%)	1.5±0.2		DBP vs. fish intake	-0.105	<0.05
			Fish <1d/wk (%)	96.9	<0.0001[a]	**Women:**			TC vs. fish intake	-0.161	<0.05
			Meat>2d/wk (%)	40.6		BMI[2] Dar es Salaam	30.4±2.3	<0.05[a]	LDLC vs. fish intake	-0.233	<0.05
			Green vegetable>2d/wk (%)	40.6		BMI[2] Handeni	27.3±2.3		Hb$_{A1c}$ vs. fish intake	-0.152	<0.05
			Coconut milk>2d/wk (%)	3.1		BMI[2] Monduli	21.5±1.9		DHA vs. fish intake	+0.347	=0.0054
			Fruits>2d/wk (%)	3.1					-3/ɷ-6 ratio vs. fish intake	+0.250	<0.05
			Whole milk>400mls/d (%)	53.1	<0.0001[a]						

Table 10 (continue): Characteristics of eight dietary assessments and health status surveys on non-communicable disease (NCD) risk factors.

Authors[Ref.] (year) country	N	Study groups location (n)	Dietary intake outcomes			Health status outcomes			Relationship between dietary variables and NCD risk factors		
			Variable	Value or prevalence	P value	Variable	Value or prevalence	p	(Study groups) NCD risk factor vs. dietary variable	r	P value
Njelekela et al[9] (2005) Tanzania	105	**Men:**	**Men:**			**Men:**					
		Urban Dar es Salaam (n=22)				**Dar es Salaam**					
						SBP^2	141.5±6.1	<0.05[a]			
						DBP^2	82.1±5.0	<0.05[a]			
						BMI^1	27.3±1.1	<0.05[a]			
						TC^4	5.1±0.3	<0.05[c]			
						$LDLC^5$	3.0±0.3	<0.05[c]			
						Hb_{A1C}^6	5.4±0.2	<0.05[c]			
		Rural Handeni (n=21)				**Handeni**					
						SBP^2	125.0±4.1				
						DBP^2	74.1±2.4				
						BMI^1	23.1±0.9				
						TC^4	3.7±0.2				
						$LDLC^5$	1.9±0.4				
						Hb_{A1C}^6	4.7±0.2				
		Pastoral Monduli (n=21)				**Monduli**					
						SBP^2	117.7±3.2				
						DBP^2	66.9±2.3				
						BMI^1	21.2±0.8				
						TC^4	4.7±0.4	<0.05[c]			
						$LDLC^5$	3.1±0.3	<0.05[c]			
						Hb_{A1C}^6	5.7±0.1	<0.05[c]			

Table 10 (continue): Characteristics of eight dietary assessments and health status surveys on non-communicable disease (NCD) risk factors.

Authors[Ref.] (year) country	N	Study groups location (n)	Dietary intake outcomes			Health status outcomes			Relationship between dietary variables and NCD risk factors		
			Variable	Value or pre-valence	P value	Variable	Value or pre-valence	p	(Study groups) NCD risk factor vs. dietary variable	r	P value
Pauletto et al[61] (1995-1996) Tanzania	1263	**Fish diet:** Lupingo village (n=618)	**Average daily calorie intake:**[†] Calories (kcal) CCH (%) Protein (%) Fats (%) Alcoholic beverage (L) Salt intake (g/d)	2196 70 18 12 0.75 4.4		**Fish diet:** SDP DBP Hypertension[7] (%) Borderline- hypertension[7] (%) Overweight[9] (%) TC TG HDL-C Cholesterol>5.18 mmol/L (%) TG>1.71 mmol/L (%) Lp(a) Total ω-6 (%) AA (%) Total ω-3 (%) EPA (%) DHA (%) n-3/n-6 ratio EPA/AA ratio DHA/AA ratio	122.6±17.5 71.8±9.2 2.8 9.7 4.5 3.53±1.04 0.92±0.64 0.86±0.32 5.5 5.0 201±213 25.8±4.8 9.7±2.7 9.7±2.9 2.3±1.3 5.7±1.6 0.39±0.13 0.24±0.08 0.61±0.14	<0.0001[a] <0.0001[a] <0.0001[a] <0.0001[a] <0.05[a] <0.0001[a] <0.0001[a] <0.0001[a] <0.0001[a] <0.0001[a] <0.0001[a] <0.001[a] <0.005[a] <0.001[a] <0.001[a] <0.001[a] <0.001[a] <0.001[a] <0.001[a]			

Table 10 (continue): Characteristics of eight dietary assessments and health status surveys on non-communicable disease (NCD) risk factors.

Authors[Ref] (year) country	N	Study groups location (n)	Dietary intake outcomes			Health status outcomes			Relationship between dietary variables and NCD risk factors		
			Variable	Value or prevalence	P value	Variable	Value or prevalence	p	(Study groups) NCD risk factor vs. dietary variable	r	P value
Pauletto et al[61] (1995-1996) Tanzania	1263	**Vegetarian diet:** Madilu village (n=645)	**Average daily calorie intake:**[†]			**Vegetarian diet:**					
			Calories (kcal)	2109		SDP	132.6±17.5				
			CCH (%)	82		DBP	76.5±9.6				
			Protein (%)	11		Hypertension[7] (%)	16.4				
			Fats (%)	7		Borderline- hypertension[7] (%)	22.3				
			Alcoholic beverage (L)	1.2		Overweight[9] (%)	2.3				
			Salt intake (g/d)	4.0		TC	4.10±1.04				
						TG	1.31±0.64				
						HDL-C	0.97±0.31				
						Cholesterol>5.18 mmol/L (%)	14.0				
						TG>1.71 mmol/L (%)	22.6				
						Lp(a)	321±212				
						Total ω-6 (%)	33.1±5.5				
						AA (%)	8.3±2.0				
						Total ω-3 (%)	3.5±1.2				
						EPA (%)	0.7±0.2				
						DHA (%)	1.5±1.1				
						n-3/n-6 ratio	0.11±0.04				
						EPA/AA ratio	0.09±0.03				
						DHA/AA ratio	0.18±0.12				

Table 10 (continue): Characteristics of eight dietary assessments and health status surveys on non-communicable disease (NCD) risk factors.

Authors[Ref.] (year) country	N	Study groups location (n)	Dietary intake outcomes			Health status outcomes			Relationship between dietary variables and NCD risk factors		
			Variable	Value or prevalence	P value	Variable	Value or prevalence	p	(Study groups) NCD risk factor vs. dietary variable	r	P value
Winmicki et al[90] (2002) Tanzania	68	**Women – fish diet:** Lapingu village (n=162)	**Daily calorie intake:** **Fish diet: all sub-groups** Maize (%) Beans (%) Spinach (%) Potatoes (%) Rice (%) Alcoholic beverage (%) Cassava (%) Fish (%) Calories (%)	14 (60-120g) 7 (40-60g) 0.4 (20-40g) 2 (40-60g) 6 (30-50g) 9 (650-850 ml) 38 (150-350g) 23 (300-600g) 100 (2196kcal)		**Women – fish diet:** Body fat (mm) Insulin (U/ml)	27.3±0.6 6.5±0.6	=0.018[d] =0.011[d]	**All sub-groups:** Plasma leptin levels vs. type of diet	NR	<0.001
						Women – vegetarian diet : Body fat (mm) Insulin (U/ml)	24.8±0.5 9.0±0.5		**Men– all sub-groups:** Plasma leptin levels vs. type of diet	0.15	=0.041
		Men– fish diet: Lapingu village (n=117)	**Women:** Alcoholic consumption (g/d) **Men:** Alcoholic consumption (g/d)	0.8±0.1 1.7±0.2		**Men– fish diet:** Body fat (mm) Insulin (U/ml)	20.2±0.6 7.6±0.6	NS NS	**Women – vegetarian diet:** Plasma leptin level vs. alcohol consumption	NR	<0.001
		Women – vegetarian diet: Madilu village (n=199)	**Daily calorie intake:** **Vegetarian diet: all sub-groups** Maize (%) Beans (%) Spinach (%) Potatoes (%) Rice (%) Alcoholic beverage Cassava (%) Fish (%) Calories (%)	42 (150-350g) 15 (70-140g) 1 (60-100g) 6 (100-200g) 17 (80-120g) 13 (600-1400ml) 6 (100-300g) negligible 100 (2109kcal)		**Men– vegetarian diet:** Body fat (mm) Insulin (U/ml)	18.5±0.6 8.4±0.7	NS NS	**Men– vegetarian diet:** Plasma leptin level vs. alcohol consumption	NR	=0.025
		Men– vegetarian diet: Madilu village (n=130)	**Women:** Alcoholic beverage intake (g/d) **Men:** Alcoholic beverage intake (g/d)	2.1±0.1 2.8±0.2	<0.001[a] <0.001[a]						

Table 10 (continue): Characteristics of eight dietary assessments and health status surveys on non-communicable disease (NCD) risk factors.

Authors[Ref] (year) country	N	Study groups location (n)	Dietary intake outcomes			Health status outcomes			Relationship between dietary variables and NCD risk factors		
			Variable	Value or prevalence	P value	Variable	Value or prevalence	p	(Study groups) NCD risk factor vs. dietary variable	r	P value
Pavan et al[62] (1997) Tanzania and Uganda; Amazon; Italy	1110	**Africa:**							**All sub-groups:**		
		rural Lugarawa district of Tanzania (n=232)	Energy intake (J)	9360		SBP[7]	144±21.9	<0.0001[a]	Serum cholesterol levels correlated to frequency of eating meat (time a week);		
			Cereal and legumes (%)	41		DBP[7]	83.2±11.8	<0.0001[a]			
			Other vegetables (%)	30.5		BMI	20.4±2.7	<0.0001[a]			
			Meat (%)	0		Overweight[9] (%)	5.4	<0.0001[a]			
			Fish (%)	15					Red meat responsible for more cholesterolaemia than white meat		<0.02
		rural *Lugbara* community of Uganda (n=138)	Dairy products and oil (%)	0		Cholesterolaemia[8] women (mg/dl)	159.7±39.1	<0.001[a]			
			Saccharose (%)	0		Cholesterol level in women (%)	6 higher	<0.05[a]	**Africa:**		
			Ethanol (%)	13.5		Glycaemia (mg/dl)	91.0±22.4	<0.0001[a]	BP vs. alcohol consumption	NR	<0.03
			NaCl (%)	4	<0.0001[a]						
			Salt intake (g/d)	4					**Brazil:**		
			Alcohol consumption (g/day)	43±61.1							
		Brazil:	Energy intake (J)	9860		SBP[7]	155.4±26.8		BP vs. alcohol consumption	NR	<0.01
		Amazon region	Cereal and legumes (%)	51.5		DBP[7]	94.5±15.5	<0.0001[e]			
			Other vegetables (%)	10.9		BMI	26.1±4.1	<0.0001[a]	**Italy:**		
		In transition from rural to urban lifestyle (n=370)	Meat (%)	10.9		Overweight[9] (%) women	60.2		BP vs. alcohol consumption	NR	<0.02
			Fish (%)	0.5		Cholesterolaemia[8] (mg/dl)	185±48.6				
			Dairy products and oil (%)	10		Cholesterol level in women (%)	9.5 higher	<0.005[a]			
			Saccharose (%)	13		Glycaemia (mg/dl)	107.9±32.7				
			Ethanol (%)	3							
			NaCl (%)	10							
			Salt intake (g/d)	10							
			Alcohol consumption (g/day)	9.6±33.4							

Table 10 (continue): Characteristics of eight dietary assessments and health status surveys on non-communicable disease (NCD) risk factors.

Authors[Ref.] (year) country	N	Study groups location (n)	Dietary intake outcomes			Health status outcomes			Relationship between dietary variables and NCD risk factors		
			Variable	Value or prevalence	P value	Variable	Value or prevalence	p	(Study groups) NCD risk factor vs. dietary variable	r	P value
Pavan et al[52] (1997) Tanzania and Uganda; Amazon; Italy	1110	**Italy:** Venice and Veneto urban/ industrialized area (n=370)	Energy intake (J) Cereal and legumes (%) Other vegetables (%) Meat (%) Fish (%) Dairy products and oil (%) Saccharose (%) Ethanol (%) NaCl (%) Salt intake (g/d) Alcohol consumption (g/day)	10750 36.3 10.2 11.5 1.7 23.7 6 10.6 11 10 38.5±45.2		SBP[7] DBP[7] BMI Overweight[9] (%) women Cholesterolaemia[8] (mg/dl) Cholesterol level in women (%) Glycaemia (mg/dl)	159.7±22.9 94.7±11.6 26.8±4.1 68.3 226.7±43.6 4.3 higher 102.8±22.6	<0.0001[a] <0.05[a]			

Table 10 (continue): Characteristics of eight dietary assessments and health status surveys on non-communicable disease (NCD) risk factors.

Authors[Ref.] (year) country	N	Study groups location (n)	Dietary intake outcomes			Health status outcomes			Relationship between dietary variables and NCD risk factors (Study groups) NCD risk factor vs. dietary variable	r	P value
			Variable	Value or prevalence	P value	Variable	Value or prevalence	p			
Mosha[33] (2003) Tanzania	976	**Morogoro district** **Women:**	**Food frequency intake:** (person/day)						**All sub-groups of obese subjects:**		
		Farmers (n=236)	Obese[10] Normal[10]	2.18±0.68 2.31±0.86		Obesity[10]	27.9%		BMI vs. feeding frequency	0.4600	<0.001
		Medium and high-level business women (n=200)	Obese[10] Normal[10]	3.80±0.63 3.80±0.63		Obesity[10]	54%				
		Civil servants (n=246)	Obese[10] Normal[10]	3.75±0.70 3.61±0.83		Obesity[10]	65%				
		House women/wives (n=29)	Obese[10] Normal[10]	3.01±0.63 2.84±0.58		Obesity[10]	49%				
		Overall (n=976)	Obese[10] Normal[10]	3.19±0.66 3.14±0.73		Obesity[10]	49%				

Results presented as: means ± standard deviation (SD) in Njelekela et al[46-48], Pauletto et al[49], Pavan et al[52], Mosha[33]
Results presented as standard error of the mean (SEM) in Njelekela et al[49], Winnicki et al[50]

Abbreviations: BP = blood pressure (mmHg); SBP = systolic blood pressure (mmHg); DBP = diastolic blood pressure (mmHg); Na = natrium; K = potassium; Na/K ratio = sodium/potassium ratio; TC = serum total cholesterol (mmol/L); HDLC = high density lipoprotein cholesterol (mmol/L); TC/HDLC = ratio of total cholesterol/HDL cholesterol; LDLC = low-density lipoprotein cholesterol (mmol/L); Hb$_{A1C}$ = haemoglobin A$_{1C}$ (%); TG = triglycerides (mmol/L); Lp(a) = lipoprotein(a) (mg/L); BMI = body mass index [body weight in kg/(height in m)^2]; AA = arachidonic acid; EPA = eicosapentaenoic acid; DHA = docosahexaenoic acid; Total ω-3 FA = total omega three fatty acids; Total ω-6 FA = Total omega six fatty acids; REE = resting energy expenditure (kcal/min/kg); d/wk = number of days per week; CCH = complex carbohydrate (i.e. maize and rice); NR = not reported
^{a}intake of food frequency was coded as either ≥3days/week or <3days/week;
^{1}Overweight = BMI ≥25 kg/m^2 and <30; obese = BMI ≥30 kg/m^2, cholesterol ≥5.2 mmol/L; ^{4}HighTC: serum cholesterol >6.2 mmol/L; Hypertriglyceridemia: serum triglyceride level above 5.2 mmol/L; ^{5}Dyslipidemia: (TC-HDLC/HDLC>5]; high HDLC: LDLC ≥4.1 mmol/L; low HDLC: HDLC <9mmol/L, ^{6}Long-standing hyperglycemia: HBA1c >7%; ^{7}Normal BP (SB<140, or DB<90 mmHg); borderline BP (SB]140-159 or DB 90-94 mmHg or both); hypertensive (SB≥160 or DB≥95 mmHg or both); ^{8}Hypercholesterolaemia: cholesterol>240 mg/dl; ^{9}Overweight: men= BMI ≥27 kg/m^2 and women = BMI ≥25 kg/m^2; ^{10}Obesity: BMI ≥25 kg/m^2; normal: BMI 18.5-24.9 kg/m^2
^{a}significantly different vs. all comparison groups; ^{b}significantly different vs. pastoral comparison group; ^{c}significantly different vs. rural comparison group; ^{d}different from female VD comparison group; ^{e}significantly different versus Africa group; ^{f}significantly different versus Brazil group; ^{g}significantly different for those eating fish more than 2 days a week

5.6. References

1. Lore W. Epidemiology of cardiovascular diseases in Africa with special reference to Kenya: An overview. *East Afr Med J.* June 1993; 70(6):357-361.
2. Yamori M. Preliminary report of cardiac study: Cross-sectional multicenter study on dietary factors of cardiovascular diseases. CARDIAC Study Group. *Clin Exp Hypertens A.* 1989; 11(5-6):957-972.
3. Edwards R, Unwin N, Mugusi F, et al. Hypertension prevalence and care in an urban and rural area of Tanzania. *J Hypertens.* February 2000; 18(2):145-152.
4. Swai A, McLarty D, Kitange H, et al. Low prevalence of risk factors for coronary heart disease in rural Tanzania. *Int J Epidemiol.* August 1993; 22(4):651-659.
5. Mufunda J, Chatora R, Ndambakuwa Y, et al. Emerging non-communicable disease epidemic in Africa: Preventive measures from the WHO Regional Office for Africa. *Ethn Dis.* 2006; 16(2):521-526.
6. Lasky B, Becerra E, Boto W, Otim M, Ntambi J. Obesity and gender differences in the risk of type 2 diabetes mellitus in Uganda. *Nutrition.* May 2002; 18(5):417-421.
7. Walker R, McLarty D, Kitange H, et al. Stroke mortality in urban and rural Tanzania. *Lancet.* May 2000; 355(9216):1684-1687.
8. Jablonski-Cohen M, Kosgei R, Rerimoi A, Mamlin J. The emerging problem of coronary heart disease in Kenya. *East Afr Med J.* June 2003; 80(6):293-297.
9. Wild S, Roglic G, Green A, Sicree R, King H. Global prevalence of diabetes: Estimates for 2000 and projections for 2030. *Diabetes*

Care. 2004; 27:1047-1053.

10. World Health Organization. *Diet, physical activity and health.* Geneva: World Health Organization; 27 March 2002.

11. Poulter N, Khaw K, Hopwood B, et al. The Kenyan Luo migration study: Observations on the initiation of a rise in blood pressure. *BMJ.* April 1990; 300(6730):967 972.

12. Shaper A, Leonard P, Jones K, Jones M. Environmental effects on the body build, blood pressure and blood chemistry of nomadic warriors serving in the army in Kenya. *East Afr Med J.* May 1969; 46(5):282-289.

13. Mugambi M, Little M. Blood pressure in nomadic Turkana pastoralists. *East Afr Med J.* December 1983; 60(12):863-869.

14. Trowell H. From normotension to hypertension in Kenyans and Ugandans 1928-1978. *East Afr Med J.* March 1980; 57(3):167-173.

15. Mtabaji J, Nara Y, Moriguchi Y, Yamori Y. Diet and hypertension in Tanzania. *J Cardiovasc Pharmacol.* 1990; 16(Suppl 8):3-5.

16. Yamori Y. WHO CARDIAC Study - its experimental background and progress report. *J UOEH.* March 1989; 11(Suppl):30-38.

17. Njelekela M, Ikeda K, Mtabaj J, Yamori M. Hypertension and its associated factors in Tanzania: A result of rapid urbanization. *J Hypertens.* Feb 2004; 22(Suppl 1):103.

18. Popkin B. An overview of the nutrition transition and its health implications: The Bellagio meeting. *Public Health Nutr.* 2002; 5:93-103.

19. Drewnowski A, Popkin B. The nutrition transition: New trends in the global diet. *Nutr Rev.* 1997; 55(2):31-43.

20. Lang T. Diet, health and globalization: Five key questions. *Proc Nutr Soc.* 1999; 58(2):335-343.

21. World Health Organization. *Diet, nutrition and the prevention of chronic diseases. Report of a Joint WHO/FAO Expert Consultation.* Geneva 2003.

22. Raschke V, Oltersdorf U, Elmadfa I, Wahlqvist M, Cheema B, Kouris-Blazos A. Content of a novel online collection of traditional East African food habits (1930s - 1960s): Data collected by the *Max-Planck Nutrition Research Unit*, Bumbuli, Tanzania. *Asia Pac J Clin Nutr.* 2007; 16(1):140-151.

23. Unwin N, Mugusi F, Aspray T, et al. Tackling the emerging pandemic of non-communicable diseases in sub-Saharan Africa: The essential NCD health intervention project. *Public Health.* May 1999; 113(3):141-146.

24. Raschke V, Oltersdorf U, Elmadfa I, Wahlqvist M, Kouris-Blazos A, Cheema B. The need for an online collection of traditional African food habits. *Afr J Food Agr Nutr Dev.* 2007; In Press.

25. Shaper A, Jones K. Serum-Cholesterol, diet and coronary heart diseas in Africans and Asians in Uganda. *Lancet.* October 1959; 10(2):534-537.

26. Barnicot N, Bennett F, Woodburn J, Pilkington T, Antonis A. Blood pressure and serum cholesterol in the Hadza of Tanzania. *Hum Biol.* February 1972; 44(1):87-116.

27. Poulter N, Khaw K, Hopwood B, et al. Blood pressure and associated factors in a rural Kenyan community. *J Hypertens.* 1984; 6:810-813.

28. Poulter N, Khaw K, Hopwood B, Mugambi M, Peart W, Sever P. Salt and blood pressure in various populations. *J Cardiovasc Pharmacol.* 1984; 6(Suppl 1):197-203.

29. World Health Organization Collaborating Centre on Primary

Prevention of Cardiovascular Disease Unit. *WHO CARDIAC study protocol and manual of operations.* Izumo, Geneva: WHO Colaborative Centre on Primary Prevention of Cardiovascular Diseases and WHO; 1986.

30. McLarty D, Swai A, Kitange H, et al. Prevalence of diabetes and impaired glucose tolerance in rural Tanzania. *Lancet.* Apr 1989; 1(8643):871-875.

31. Shaper A, Jones K, Jones M, Kyobe J. Serum lipids in three nomadic tribes of Northern Kenya. *Am J Clin Nutr.* September 1963; 13:135-146.

32. Katsivo M, Apeagyie F. Hypertension in Kitui district: A comparative study between urban and rural populations. *East Afr Med J.* July 1991; 68(7):531-538.

33. Kitange H, Swai A, Masuki G, Kilima P, Alberti K, McLarty D. Coronary heart disease risk factors in sub-Saharan Africa: Studies in Tanzanian adolescents. *J Epidemiol Community Health.* August 1993; 47(4):303-307.

34. McLarty D, Unwin N, Kitange H, Alberti K. Diabetes mellitus as a cause of death in sub-Saharan Africa: Results of a community-based study in Tanzania. *Diabet Med.* November 1996; 13(11):990-994.

35. Mgonda Y, Ramaiya K, Swai A, McLarty D, Alberti K. Insulin resistance and hypertension in non-obese Africans in Tanzania. *Hypertension.* January 1998; 31(1):114-118.

36. Marcovina S, Kennedy H, Bittolo Bon G, et al. Fish intake, independent of apo(a) size, accounts for lower plasma lipoprotein(a) levels in Bantu fishermen of Tanzania: The Lugalawa Study. *Arterioscler Thromb Vasc Biol.* May 1999; 19(5):1250 1256.

37. Bovet P, Ross A, Gervasoni J, et al. Distribution of blood pressure,

body mass index and smoking habits in the urban population of Dar es Salaam, Tanzania, and associations with socioeconomic status. *Int J Epidemiol.* Feb 2002; 31(1):240-247.

38. Kuga S, Njelekela M, Noguchi T, et al. Prevalence of overweight and hypertension in Tanzania: Special emphasis on resting energy expenditure and leptin. *Clin Exp Pharmacol Physiol.* October 2002; 29(S4):23-26.

39. Njelekela M, Negishi H, Nara Y, et al. Obesity and lipid profiles in middle aged men and women in Tanzania. *East Afr Med J.* Feb 2002; 79(2):58 64.

40. Njelekela M, Kuga S, Hiraoka J, et al. Determinants of hyperleptinaemia in an African population. *East Afr Med J.* April 2003; 80(4):195-199.

41. Villamor E, Msamanga G, Urassa W, et al. Trends in obesity, underweight, and wasting among women attending prenatal clinics in urban Tanzania, 1995-2004. *Am J Clin Nutr.* June 2006; 83(6):1387-1394.

42. Mazengo M, Simell O, Lukmanji Z, Shirima R, Karvetti R. Food consumption in rural and urban Tanzania. *Acta Trop.* December 1997; 68(3):313-326.

43. Kinabo J, Mnkeni A, Nyaruhucha C, Msuya J, Haug A, Ishengoma J. Feeding frequency and nutrient content of foods commonly consumed in the Iringa and Morogoro regions in Tanzania. *Int J Food Sci Nutr.* Feb-Mar 2006; 57(1-2):9-17.

44. Mann GV, Schäffer RS, Anderson RS, Sandstead HH. Cardiovascular disease in the Masai. *J Atheroscler Res.* 1964; 4:289-312.

45. Mtabaji J, Nara Y, Yamori Y. The cardiac study in Tanzania: Salt

intake in the causation and treatment of hypertension. *J Hum Hypertens.* April 1990; 4(2):80-81.

46. Njelekela M, Negishi H, Nara Y, et al. Cardiovascular risk factors in Tanzania: A revisit. *Acta Trop.* June 2001; 79(3):231-239.

47. Njelekela M, Kuga S, Nara Y, et al. Prevalence of obesity and dyslipidemia in middle-aged men and women in Tanzania, Africa: Relationship with resting energy expenditure and dietary factors. *J Nutr Sci Vitaminol (Tokyo).* October 2002; 48(5):352-358.

48. Njelekela M, Sato T, Nara Y, et al. Nutritional variation and cardiovascular risk factors in Tanzania: Rural-urban difference. *S Afr Med J.* April 2003; 93(4):295-299.

49. Njelekela M, Ikeda K, Mtabaji J, Yamori Y. Dietary habits, plasma polyunsaturated fatty acids and selected coronary disease risk factors in Tanzania. *East Afr Med J.* Nov. 2005; 82(11):572-578.

50. Winnicki M, Somers V, Accurso V, et al. Fish-rich diet, leptin, and body mass. *Circulation.* July 16 2002; 106(3):289-291.

51. Pauletto P, Puato M, Caroli M, et al. Blood pressure and atherogenic lipoprotein profiles of fish-diet and vegetarian villagers in Tanzania: The Lugalawa study. *Lancet.*Sept 1996; 348(9030):784-788.

52. Pavan L, Casiglia E, Pauletto P, et al. Blood pressure, serum cholesterol and nutritional state in Tanzania and in the Amazon: Comparison with an Italian population. *J Hypertens.* October 1997; 15(10):1083-1090.

53. Mosha T. Prevalence of obesity and chronic energy deficiency (CED) among females in Morogoro district, Tanzania. *Ecol Food Nutr.* 2003; 42(1):37-67.

54. Bunker C, Ukoli F, Okoro F, et al. Correlates of serum lipids in a lean black population. *Atherosclerosis.* June 1996; 123(1 2):215-

225.

55. Snook J, Park S, Williams G, Tsai Y-H, Lee N. Effect of synthetic triglycerides of myristic, palmitic, and stearic acid on serum lipoprotein metabolism. *Eur J Clin Nutr.* August 1999; 53(8):597-605.

56. St-Onge M, Farnworth E, Jones P. Consumption of fermented and nonfermented dairy products: Effects on cholesterol concentrations and metabolism. *Am J Clin Nutr.* March 2000; 71(3):674-681.

57. Johns T, Mahunnah R, Sanaya P, Chapman L, Ticktin T. Saponins and phenolic content in plant dietary additives of a traditional subsistence community, the Batemi of Ngorongoro District, Tanzania. *J Ethnopharmacol.* July 1999; 66(1):1-10.

58. Johns T, Nagarajan M, Parkipuny M, Jones P. Maasai Gummivory: Implications for Paleolithic Diets and Contemporary Health. *Curr Anthropol.* 2000; 41:453-459.

59. Mendez M, Monteiro C, Popkin B. Overweight exceeds underweight among women in most developing countries. *Am J Clin Nutr.* March 2005; 81(3):714-721.

60. Drummond S, Crombie N, Kirk T. A critique of the effects of snacking on body weight status. *Eur J Clin Nutr.* December 1996; 50(12):779-783.

61. Poulter N, Khaw K, Hopwood B, et al. Blood pressure and its correlates in an African tribe in urban and rural environments. *J Epidemiol Community Health.* 1984; 38(3):181-185.

62. Hodge A, Dowse G, Erasmus R, et al. Serum lipids and modernization in coastal and highland Papua New Guinea. *Am J Epidemiol.* December 1996; 144(12):1129-1142.

63. McMurry M, Cerqueira M, Connor S, Connor W. Changes in lipid

and lipoprotein levels and body weight in Tarahumara Indians after consumption of an affluent diet. *N Engl J Med.* December 1991; 325(24):1704-1708.

64. Knowler W, Pettitt D, Saad M, Bennett P. Diabetes mellitus in the Pima Indians: Incidence, risk factors and pathogenesis. *Diabetes Metab Rev.* February 1990; 6(1):1 27.

65. Leaf A. Cardiovascular effects of fish oils. Beyond the platelet. *Circulation.* August 1990; 82(2):624-628.

66. Tremoli E, Maderna P, Marangoni F, et al. Prolonged inhibition of platelet aggregation after n-3 fatty acid ethyl ester ingestion by healthy volunteers. *Am J Clin Nutr.* March 1995; 16(3):607-613.

67. Bang H, Dyerberg J, Sinclair H. The composition of the Eskimo food in north western Greenland. *Am J Clin Nutr.* December 1980; 33(12):2657-2661.

68. Kromhout D, Bosschieter E, de Lezenne Coulander C. The inverse relation between fish consumption and 20-year mortality from coronary heart disease. *N Engl J Med.* May 1985; 312(9):1205-1209.

69. Kromhout D, Keys A, Aravanis C, et al. Food consumption patterns in the 1960s in seven countries. *Am J Clin Nutr.* May 1989; 49(5):889-894.

6. UTILIZATION AND FUTURE APPLICATIONS FOR AN ONLINE COLLECTION OF TRADITIONAL AFRICAN FOOD HABITS

6.1. Background and objectives for the web site evaluation

The challenges facing the people of sub-Saharan Africa in the 21st century are myriad. While epidemics of malnutrition and HIV/AIDS continue unabated,[1] non-communicable diseases (NCDs) have also surged to epidemic levels over the past few decades. Current statistics from the World Health Organization reveal that nearly 80% of deaths attributable to NCDs occur in developing countries, including those of sub-Saharan Africa.[2]

Amongst the contributing factors to the upsurge of NCDs has been the deleterious shift from traditional food habits to the processed and packaged food products of transnational corporations.[3, 4] These products tend to be high in saturated fat, trans fatty acids, and food preservatives; and low in dietary fibre, vital nutrients, and phytochemicals when compared to basic dietary guidelines.[5-7] The shift from traditional food habits to the food products of the transnational corporations has been dubbed the *nutrition transition*, and has been directly implicated in the rise of NCDs and risk factors including obesity, hypertension, abnormal lipid profiles, type 2 diabetes, cardiovascular disease (CVD), and certain cancers throughout sub-Saharan Africa.[8-10] Further, the recent epidemics of NCDs have created a polarized and protracted double burden of disease as the effects of the *nutrition transition* add to the existing infectious and malnutrition disease burden of the region.[9, 10]

Throughout history, external influences have brought about changes in African food habits, and this has perhaps never been more apparent than the present day (See Chapter 4). As the people of sub-Saharan Africa continue to experience a *nutrition transition* toward the manufactured food products of the transnational corporations, knowledge of traditional African foods and food habits will continue to be lost. This outcome is distressing, especially given the fact that traditional African food habits are extremely healthy and may indeed provide a clear solution to current NCD epidemics sweeping the continent.

According to a survey conducted at the 18th International Congress on Nutrition (ICN) in Durban, South Africa, 2005,[11] 84% of experts in the nutritional sciences (n=92) agreed that traditional African food habits were superior to the *globalized* food culture currently underpinning the *nutrition transition*. Further, these experts believed that the knowledge of traditional African food habits is being lost (80%) and that there is a vital need for investigation and documentation to preserve this knowledge.[11]

To address this existing gap we created an online collection, the first of its kind, to preserve and disseminate knowledge of traditional African foods and food habits.[12] The first available data for our online collection were amalgamated by reviewing a series of observational studies collated by the *Max-Planck Nutrition Research Unit*, formerly located in Bumbuli, Tanzania. This series of studies is titled the *Oltersdorf Collection*, after Professor Ulrich Oltersdorf, who was involved in some of the research at the *Max-Planck Nutrition Research Unit*. Professor Ulrich Oltersdorf contributed significantly to this project by making the collection available to our investigative team.[13]

At present, the online collection (which was available at: http://www.healthyeatingclub.com/Africa/ and has been transferred to http://www.drverena.com)[12] presents the first empirical evidence of traditional foods and food habits of East Africa (i.e. Tanzania, including

Zanzibar Island and Pemba Island, Kenya, and Uganda) from the 1930s to the1960s. [14] The collection[12] also provides information regarding several recent research projects being conducted by indigenous African scientists which provide information on traditional African food habits and their health benefits.

The majority of interviewees (92%) surveyed at the recent ICN in Durban, South Africa, stated they would make use of our online information for educational and research purposes if this information was made available.[11] Our collection has been available via the worldwide web since May 2006. The utilization, however, of the content within the collection has not been investigated to date. Therefore, the purpose of the present investigation was to evaluate the overall utilization of our online information system in order to determine the most requested PDF documents and web pages, and to discuss the potential uses of our online collection with regards to future application.

6.2. Methods

6.2.1. The online collection

The Oltersdorf Collection (1930s to 1960s)

The focal point of our online information system is the *Oltersdorf Collection,* 75 observational reports of nutrition-related investigations collected by the *Max-Planck Nutrition Research Unit*, formerly located in Bumbuli, Tanzania. The *Oltersdorf Collection*[14] has been systematically reviewed to amalgamate and present the data related to traditional foods and food habits of East Africa (i.e. Tanzania, Zanzibar Island and Pemba Island, Kenya, and Uganda) from the 1930s to 1960s. The entire collection of documents has been scanned and converted into PDF files, which are now available for free download via the web site on African food habits.[12]

Recent research projects

In addition to the *Oltersdorf Collection*, our web site provides recent data and publications concerning traditional African food habits from indigenous African scientists within the nutritional sciences. New information will continue to be added as more information is compiled and put forth.

6.2.2. Web site structure

A detailed description of the development of the online collection structure and the allocation of country specific web page topics is presented in Chapter 2, and published by the *Asia-Pacific Journal of Clinical Nutrition*.[14]

6.2.3. Measurement of the web site traffic

The online collection (web site) traffic was recorded by Server 101, a web host for the *Healthy Eating Club (HEC)* account through which the online collection has been made available.[12] Server 101 provides a weekly (Saturday to Saturday) web site statistic report, analyzed with Analog 5.32 software. The web host (Server 101) provides statistics on: (1) the directory report, which provides weekly traffic on the collection,[12] and (2) weekly web page requests, which lists files, including PDF documents, with at least 3 requests per week.

6.2.4. Analysis of web site statistics

The web site traffic was recorded each week in a Microsoft Excel spreadsheet over a time period of 31 weeks, from the time the collection went online in March 2006 (12th week of 2006; 25.03.2006) until October

2006 (42nd week of 2006; 22.10.2006). The web site data was analyzed to determine the following:

1. Total visits to the online collection per week over 31 weeks
2. The ten most requested PDF files and web pages, excluding the index page and the overview content page, over 31 weeks
3. The top most requested region specific web pages of the four country directories (i.e. Tanzania, Zanzibar Island and Pemba Island, Kenya and Uganda) over 31 weeks
4. The most requested web page topics over 31 weeks.

6.2.5. Limitations

The request report provided by Server 101, only reports on the number of requested PDF files, html and php pages, requested at least more than three times over a time period of seven days. In addition, the number of requested images per week, of which the web site holds currently 115, is not reported within the request report.

6.3. Results

6.3.1. Total visits to the online collection

Over 31 weeks 91687 visits were recorded to the African food habits web site. Total visits to the online collection per week[12] over a time period of 31 weeks are presented in Figure 10. The total visits increased 14% from the first to second week. From week 1 to week 20, total visits declined by approximately 49%. From week 20 to week 28, the number of total web site visits increased by 55%. Overall, the number of visits increased by 17% from week 1 to week 31. On average, our online collection accounted for 2958 visits per week.

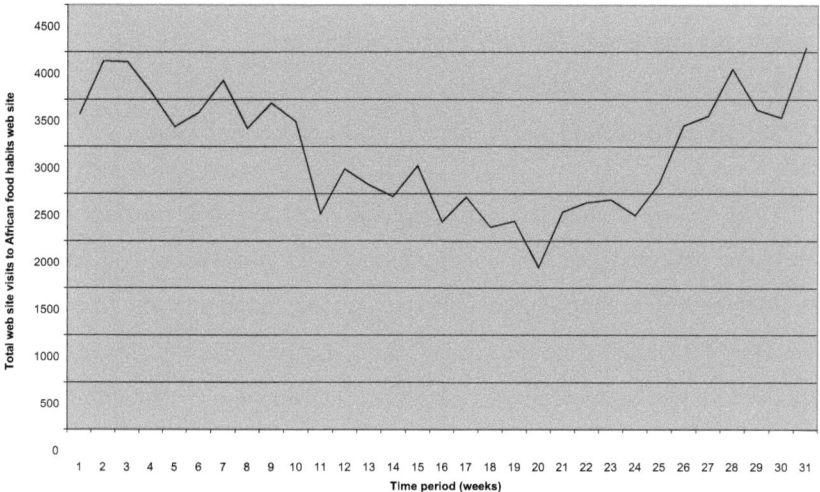

Figure 10. Total visits to the African food habits web site over 31 weeks.

6.3.2. Ten most requested PDF files and web pages

The ten most requested PDF files and web pages over 31 weeks are presented in Figure 11. In order from most popular, these included:

1. *Introduction to the PhD Project* web page
2. *Comparison of Nutrient Intake in East Africa* by Oltersdorf et al.[13], a PDF document from the *Oltersdorf Collection*
3. *Tribes of Tanganyika* by Jerrard et al.[15], a PDF document from the *Oltersdorf Collection*
4. *Maasai Traditional Foods: A Look at Maasai Diets in the Maasai Culture* by Imbumi M. et al.,[16] a PDF document of a poster presented at the ICN in Durban, South Africa, in 2005.
5. *Kenya Contents* web page
6. *Minimum Dietary Standards for East African Natives* by Raymond et al.,[17] a PDF document from the *Oltersdorf Collection*

7. *Tanzania Contents* web page
8. *Background to the PhD Project* web page
9. *UgandaContents* web page
10. *Zanzibar Island and Pemba Island Contents* web page.

The ten most requested web pages and PDF documents included two web pages that provided introductory (Rank: #1) and background (Rank: #8) information to our online collection. The three reports from the *Oltersdorf Collection* were the second, third and sixth most requested PDF files over 31 weeks, respectively. A recent investigation by Imbumi M. *et al.*[16], was the fourth most requested file on the African food habits web site

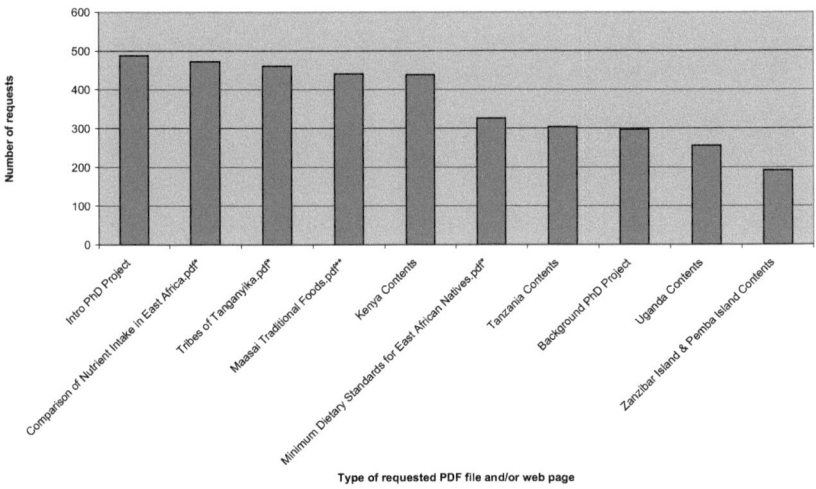

*Report from the *Oltersdorf Collection*
**Recent investigation by Imbumi *et al.*[16]

Figure 11. Most requested PDF files and/or web pages over 31 weeks.

over a time period of 31 weeks. Among the four region specific content pages (i.e. Tanzania, Zanzibar Island and Pemba Island, Kenya and

Uganda), the *Kenya contents* page was the most frequently visited web page, followed by the *content pages* for *Tanzania, Uganda and Zanzibar Island and Pemba Island*, respectively.

6.3.3. Most requested topics

The most visited topics of the four country directories over 31 weeks are presented in Figure 12. The web page entitled *Kenya Vegetables* was the most requested web page among all the country directory web pages, with 169 requests. The second most requested web page was the *Zanzibar Island and Pemba Island Diet and Dishes* followed by *Kenya Diet and Dishes*, with 138 and 117 requests, respectively.

The most visited web page topics of the four country directories combined over a time period of 31 weeks, are presented in Figure 13. *Diet and Dishes* was the most requested web page topic, followed by *Vegetables* and the *Chemical Composition of Traditional African Foods*, with 352, 239 and 172 requests, respectively.

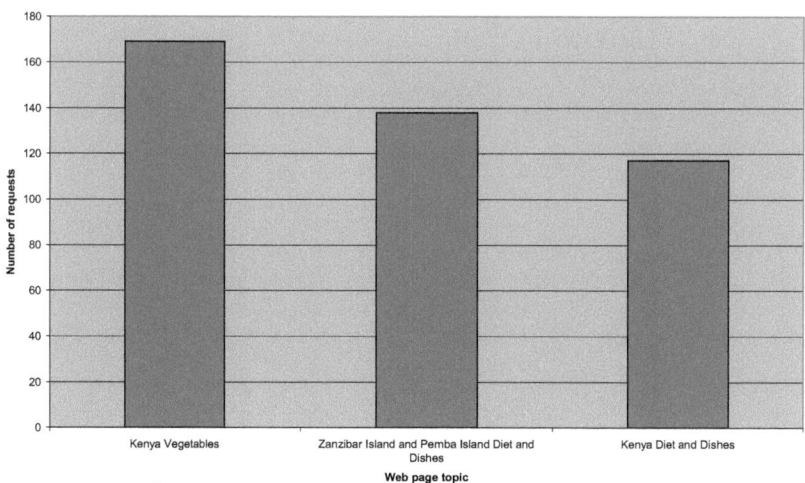

Figure 12. Most requested web pages topics among the four country directories over

31 weeks.

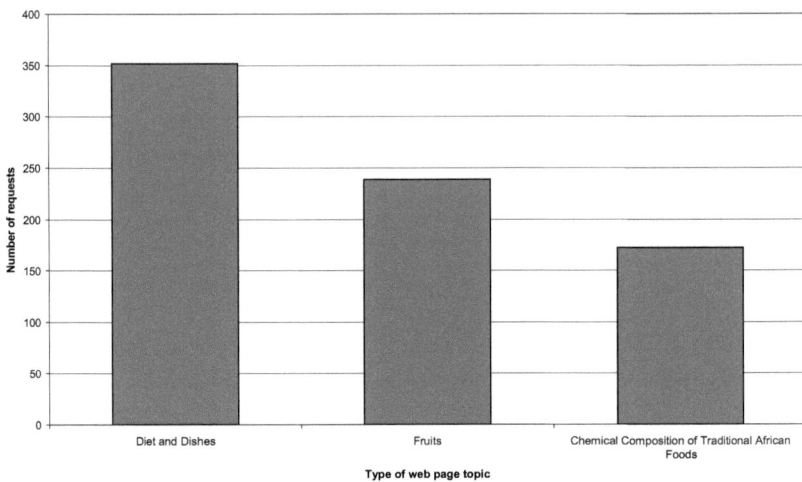

Figure 13. Most requested web pages topics over 31 weeks, combined for all countries

6.4. Discussion

Knowledge of traditional African food habits is being lost. There is clearly a need for documentation, compilation, and dissemination of this rapidly eroding wealth of information.[11] Our investigative group, including the principle investigator (V.R.) and four co-investigators in the field of nutritional sciences (U. Oltersdorf, I. Elmadfa, M.L. Wahlqist and A. Kuris-Blazos), have contributed to collating and honing knowledge of traditional foods and food habits in Africa by establishing an online information system on African food habits.[12] This online collection[12] currently presents data related to traditional foods and food habits of East Africa (i.e. Tanzania, including Zanzibar Island and Pemba Island, Kenya, and Uganda) from the 1930s to 1960s, based on a systematic review of the *Oltersdorf Collection*.[11]

To investigate the utilization of the online collection,[11] the web site traffic was analyzed. Overall, the results revealed a 17% increase in total web site visits from week 1 to week 31. On average, the online collection accounted for 2958 visits per week. Further, the results of the present investigation reveal an interest in the PhD project (i.e. Introduction and Background web pages),[12] innovative research conducted by indigenous African scientists,[16] and the *Oltersdorf Collection* (Figure 11).[11] These web pages were amongst the most requested files over the specified time period (31 weeks). These findings provide support for creating an online collection of traditional African food habits based on both historical[14] and current research.[16] This collection can potentially serve as a well rounded forum in which information on traditional African foods and food habits is presented and shared.

The *Kenya Content* page was the most visited web page among the four country directories, and this may be associated with the high number of downloads of the PDF file entitled *Maasai Traditional Foods: A Look at Maasai Diets in the Maasai Culture* by Imbumi et al.[16] and the high number of requests for the web pages *named Kenya Vegetable* and *Kenya Diet and Dishes*.

The results of the web site utilization of the online collection on African food habits revealed that *Diet and Dishes Vegetables* and *Chemical Composition of Traditional African Foods* were the most requested web page topics for the countries combined over a time period of 31 weeks. Amalgamating indigenous knowledge regarding food choices in combination with historical empirical knowledge and innovativescientific investigations of the chemical composition of foods may increase the marketability of traditional African food items.

The majority of stakeholders (92%) interviewed at the 18th ICN, in

Durban, South Africa, 2005, indicated that they would make use of such an innovative online information on traditional African foods and food habits, if available. The results of the web site utilization performed suggest that the web site is frequently accessed (2958 visits per week, on average) and the PDF files posted on the website are frequently downloaded, as four PDF files were amongst the 10 most frequently downloaded web pages. It can be hypothesized that our online collection can successfully contribute to the documentation, compilation, and dissemination of information related to traditional African food habits.[18]

6.4.1. Importance of the Oltersdorf Collection

As the *Oltersdorf Collection*[14] provides some of the first empirical evidence of traditional food habits in Africa, it has the potential to trigger more thorough investigation of traditional African food habits in the coming years, and may encourage the revelation of additional sources of traditional food knowledge from Africa. Important implications of this valuable historical collection are provided in Table 11.

Table 11. Importance of the *Oltersdorf Collection*.

The historical data set being made available online will improve:
- *Access:* providing these data, for the first time, via electronic medium
- ***Research:*** by providing a reference point for future studies conducted in these countries, research will be improved and inter-ethnic/community research can be enhanced
- *Education:* increasing awareness of this valuable data set that spans 3 decades (1930s to 1960s) and includes the earliest, fundamental nutrition and health status surveys carried out in East Africa

- ***Communication:*** the web site including e-versions of old but valuable documents may be an important medium for communication in and outside Africa
- ***Comparison:*** by providing these historical dataset for the first time, via electronic medium, the early changes in food habits (past-present) and its driving forces can be explored; Thus, comparative data on food quality and health status during periods of transition may enhance advocacy for the traditional diet amongst health care, and nutrition intervention programs throughout Africa, and perhaps globally
- ***Dietary acculturation:*** providing information on East African foods and food habits may enhance the situation of African refugees after migration as food security issues* can be identified by governmental institutions and health care providers
- ***Awareness:*** by providing information on indigenous, traditional African foods and their health benefits, African and Western countries could potentially increase the availability of particular African foods

*availability and affordability as well as identification of indigenous, traditional East African foods and food habits to enhance dietary diversity and healthy food choices after migration.

6.4.2. Future capacity of the African food habits web site

Further documentation of the knowledge of traditional African foods and food habits and their relation to health may be necessary to gain an understanding of how traditional dietary patterns could potentially abate projected trends for NCDs and malnutrition in Africa, and perhaps globally.[14, 19, 20]

In addition, reliable strategies to reduce food insecurity in rural settings of sub-Saharan Africa require the acknowledgement of Africa's rich food culture.[14] According to the International Fund for Agricultural Development (IFAD)[21] household food security should consist of "food adequacy while complying with nutrient and safety requirements and *cultural preferences*." We believe that the future objectives of the online collection should be two-fold:

1. To amalgamate current and historical research data related to traditional foods and food habits throughout all of Africa
2. To develop an online network for communication, in and outside of Africa, to develop targeted and relevant collaborative research projects.

6.4.3. Transferability of African foods and food habits

The irony of the *nutrition transition* is that people in developing countries abandon their traditional diet as 'backward' and 'poor' and are made to favor fashionable, modern western foods.[24, 25] This is generally facilitated by cultural indoctrination, scarcity of traditional foods, and abundance of less healthy, western foods.[14] However, in the West, people are looking at traditional diets such as those of the Mediterranean and East Asia, as a source of information, optimal nutrition, and health. Indeed, investigations in Okinawa Japan,[26, 27] the Mediterranean,[28, 29] and China,[30, 31] have provided robust evidence that traditional foods and food habits are inextricably linked to vitality and longevity.

In comparison to these particular food cultures, traditional African diets have not been considered important.[32, 33] However, traditional African foods and food habits are associated with a broad range of health benefits according to the latest annals of scientific inquiry.[14] Traditional African diets hold the potential to address problems such as communicable diseases,[34, 35] cardiovascular diseases,[36, 37] diabetes,[38] and micronutrient deficiencies.[39-42]

The risk-reduction potential of traditional African diets is particularly due to their foundation on dietary diversity,[34, 43] the use of traditional edible species and constituents of high nutritional value,[39, 44, 45] traditional food preparation techniques and cooking methods.[39, 46, 47] Particular health related functions of African plants, include antibiosis[48], immunostimulation,[49] anti-gout,[45] antioxidant,[50] hypoglyecemic[38] and hypolipidemic properties.[51-53]

The availability of information on traditional food habits seems even more important in relation to the use of plant species and their related health benefits in nutrition interventions.[54] The admixture of different foods in a meal can influence the bioavailability of specific components from the different foods. If viewed in terms of components: season, growing conditions, food storage, food preparation and additional factors will all influence, to a greater or lesser extent, the composition of the meal.[55]

The knowledge of traditional African foods and food habits, which has evolved and remains at the heart of this *Cradle of Civilization*, including a wide knowledge of the relationship between food, food habits and longevity, should be investigated further. At long last, realization of the potential of traditional foods and food habits and sociocultural values in addressing the *double burden* of disease is improving. Recently, several scientific research projects have been developed towards strengthening the capacity and utilization of traditional African foods and food habits, counteracting the development of growing simplifications of diets in

Africa.[12, 14, 22, 56-58] This particular progress supports the potential of traditional African foods and food habits to encourage the following developments:

1. *The improvement of disease epidemics*

Increased attention to information and knowledge of indigenous African foods and food habits by the global community may improve the current *globalized* food culture, which has largely been responsible for the global rise of NCD epidemics. Furthermore, the promotion of locally accepted, traditional African foods would allow for nutritious, inexpensive, and healthy dietary recommendations for indigenous African people living with, and at risk of, diet-related NCDs, including type 2 diabetes, obesity, and cardiovascular disease.[59] Thus, the integration of knowledge regarding the health-related benefits of local, traditional African foods and food habits in nutrition education may result in more appropriate treatments for type 2 diabetes and other NCDs than currently being offered by the transnational pharmaceutical companies.

2. *Marketing opportunities*

The health benefits associated with indigenous African species and traditional food habits may hold marketing potential. The movement of people between countries and regions may provide further opportunities to strengthen African markets, and restaurants within Africa and abroad, to the economic benefit of African migrants and refugees. At the same time, tourism to Africa may represent an increasingly important method of economically supporting local markets and restaurants built around the demand of travelers for healthy, natural foods. It is paramount that the commercialization of indigenous African food habits is to the benefit of the indigenous African population, especially the small and resource poor farmers, rather than new world professionals and transnational corporations interested in patenting natural foods. The commercialization of traditional African food habits may be facilitated by the strong promotion, particularly from academia via books and other lay publications.[60]

2. *Food security*

Many traditional African crops are nutritionally rich and adapted to low input agriculture.[32, 45, 61] Thus, they offer the potential to address serious problems of food security and undernutrition which continue to affect Africa, and other developing countries. Accordingly, traditional African crops may be transferred into other nutritionally-challenged areas of the world, with similar geoclimatic characteristics.

3. *Ecosystem stability*

Indigenous African crops are better adapted to stress and difficult environments compared to highly commercialized crops (e.g. wheat, rice, maize) and may have a key role in maintaining diversity and a more stable environment, and providing food security during severe climatic fluctuations.

In summary, the results of the web site utilization suggest that the web site on traditional African foods and food habits is frequently accessed and the PDF files posted on the web site are frequently downloaded. It can be hypothesized that our online collection can successfully contribute to the documentation, compilation, and dissemination of information related to traditional African food habits. As the *Oltersdorf Collection* provides some of the first empirical evidence of traditional food habits in Africa, it has the potential to trigger more thorough investigation of traditional African food habits in the coming years, and may encourage the discovery of additional sources of traditional food knowledge from Africa. The future expansion of the web site project will include the development of data availability of current and historical research data related to traditional foods and food habits throughout all of Africa. As well, this web site has the potential for the development of an online network for communication, in and outside of Africa, for developing targeted and relevant collaborative research projects.

6.5. Résumé

As the people of sub-Saharan Africa continue to experience a *nutrition transition*, knowledge of traditional African foods and food habits will continue to wane. This outcome is distressing, as traditional African food habits are extremely healthy and may, therefore, help to alleviate current trends related to non-communicable diseases (NCDs). To preserve and disseminate knowledge of traditional African foods and food habits, an online collection was created which has been available since May 2006, via the worldwide web. However, the utilization of the collection has not been investigated to date. Therefore, the purpose of the present investigation was to evaluate the overall utilization of our online collection over the first 31 weeks of availability, and to discuss its future applications. The online collection (web site) traffic was recorded by Server 101, which provides a weekly (Saturday to Saturday) web site statistic report, analyzed with Analog 5.32 software. Overall, the number of visits increased by 17% from week 1 to week 31. On average, our online information system accounted for 2958 visits per week. The ten most requested web pages and PDF documents included: two web pages that provided introductory and background information of the project, three reports from the *Oltersdorf Collection*, a recent investigation of the food habits of the Maasai in Kenya, conducted by indigenous African researchers Imbumi *et al.*; and four content pages listed by region (i.e. 1. Tanzania; 2. Zanzibar Island and Pemba Island; 3. Kenya; and 4. Uganda). Using the pooled data of all regions, the most popular web page topics were: 1. *Diet and Dishes*, followed by 2. *Vegetables,* and the 3. *Chemical Composition of Traditional African Foods*. Overall, our results suggest that our online collection is increasing in popularity (by 17%), and is frequently accessed for various topics and PDF files. It can be hypothesized that our online information system can successfully contribute to the documentation, compilation, and dissemination of information related to traditional African food habits. The future development of the web site project will include the expansion of data availability of current and historical research data pertaining to traditional foods and food habits throughout all of Africa. In addition, the web site will include the development of an online network of communication, in and outside of Africa, to develop targeted and relevant collaborative research projects related to traditional African foods.

6.6. References

1. World Health Organization. The World Health Report. Today's Challenges. Geneva: World Health Organization; 2003.
2. World Health Organization. Diet, physical activity and health. Geneva: World Health Organization; 27 March 2002.
3. Maletnlema T. A Tanzanian perspective on the nutrition transition and its implication for health. *Public Health Nutr.* 2002; 5(1A):163-168.
4. Popkin BM, Lu B, Zhai F. Understanding the nutrition transition: Measuring rapid dietary changes in transitional countries. *Public Health Nutr.* 2002; 5(6A):947-993.
5. Shetty P. *Diet nutrition and chronic disease*: Lessons from contrasting worlds. Vol 15-16. Chichester, UK: John, Wiley and Sons; 1997.
6. Drewnowski A, Popkin B. The nutrition transition: New trends in the global diet. *Nutr Rev.* 1997; 55(2):31-43.

7. Walker A, Segal I. Health/ill-health transition in less privileged populations: what does the future hold? *J R Coll Physicians Lond.* 1997; 31(4):392-395.
8. World Health Organization (WHO). *Globalization, diet and noncommunicable disease.* Geneva 2002.
9. Popkin B. An overview of the nutrition transition and its health implications: The Bellagio meeting. *Public Health Nutr.* 2002; 5:93-103.
10. Popkin BM. The nutrition transition and prevention of diet related disease in Asia and the Pacific. *Food Nutr Bull.* 2001; 22:S1-58.
11. Raschke V, Oltersdorf U, Elmadfa I, Wahlqvist M, Kouris-Blazos A, Cheema B. The need for an online collection of traditional African food habits. *Afr J Food Agr Nutr Dev.* 2007; In Press.
12. Raschke V. East African Food Habits On-line. In: Wahlqvist ML. *HEC Press.* Available at: http://www.healthyeatingclub.org/Africa/index.htm.
13. Oltersdorf U. Comparison of nutrient intakes in East Africa. Paper presented at: The Human Biology of Environmental Change.; 5-12 April, 1971; Blantyre, Malawi.
14. Raschke V, Oltersdorf U, Elmadfa I, Wahlqvist M, Cheema B, Kouris-Blazos A. Content of a novel online collection of traditional East African food habits (1930s - 1960s): Data collected by the *Max-Planck Nutrition Research Unit*, Bumbuli, Tanzania. *Asia Pac J Clin Nutr.* 2007; 16(1): 140-151.
15. Jerrard R. Tribes of Tanganyika, their districts, usual dietary and pursuits. Dar es Salaam: Government Printer; 1935.
16. Imbumi M, Saitabu H, Maundu P. Maasai traditional foods: A look at diets in the Maasai culture. *Ann Nutr Metab.* September 2005;

49(Suppl. 1):381.

17. Raymond W. Minimum dietary standards for East African natives. *East Afr Med J.* 1940; 17:249.

18. Wahlqvist M. Towards A New Generation of International Nutrition Science and Scientist: The Importance of Africa and Its Capacity. *J Nutr.* 2006; 136(4):1048-1049.

19. Unwin N, Setel P, Rashid S, et al. Noncommunicable diseases in sub-Saharan Africa: where do they feature in the health research agenda? *Bull World Health Organ.* 2001; 79(10):947-953.

20. Wahlqvist M. Regional food diversity and human health. *Asia Pac J Clin Nutr.* 2003; 12(3):304-308.

21. International Fund for Agricultural Development (IFAD). Draft Report of the consultation to review the adequacy of the resources available to IFAD 2000-2002. Rome: International Fund for Agricultural Development (IFAD); February 16-17 2000.

22. van der Walt R, Bezuidenhout C, Bouwman B. Initiative for Development of Indigenous Food-plants of Africa (IDIFA). Paper presented at: Cape to Cairo Safari Conference, 2005; North-West University Potchefstroom.

23. van der Walt A. *Draft workshop report: Second meeting of the Initiative for the Development of Indigenous Food-plants of Africa (IDIFA) hosted by the Faculty of Agriculture, Cairo University, Egypt, 10-11 April 2006.* Potchefstroom: Initiative for the Development of Indigenous Food-plants of Africa (IDIFA); 2006.

24. Voster H, Bourne L, Venter C, Oosthuizen W. Contribution of nutrition to the health transition in developing countries: A framework for research and intervention. *Nutr Rev.* 1999; 57(11):341-349.

25. Popkin B. Global nutrition dynamics: The world is shifting rapidly

toward a diet linked with noncommunicable diseases. *Am J Clin Nutr.* Aug 2006; 84(2):289-298.

26. Yamori Y, Miura A, Taira K. Implications from and for food cultures for cardiovascular diseases: Japanese food, particularly Okinawan diets. *Asia Pac J Clin Nutr.* 2001; 10(2):144-145.

27. Cockerham W, Yamori Y. Okinawa: An exception to the social gradient of life expectancy in Japan. *Asia Pac J Clin Nutr.* June 2001; 10(2):154-158.

28. Trichopoulou A, Critselis E. Mediterranean diet and longevity. *Eur J Cancer Prev.* October 2004; 13(5):453-456.

29. Serra-Majem L, Roman B, Estruch R. Scientific evidence of interventions using the Mediterranean diet: a systematic review. *Nutr Rev.* February 2006; 64(Suppl 1):27-47.

30. MacLennan R, Zhang A. Cuisine: The concept and its health and nutrition implications--global. *Asia Pac J Clin Nutr.* 2004; 13(2):131-135.

31. Liu X, Li Y. Epidemiological and nutritional research on prevention of cardiovascular disease in China. *Br J Nutr.* December 2000; 84(Suppl 2):199-203.

32. IPGRI. Conserving and increasing the Use of Neglected and Underutilized Crop Species. International Plant Genetic Resources Institute (IPGRI). Available at: http://www.ipgri.cgiar.org/nus/strategy.htm. Accessed November 10, 2006.

33. AFROL News. Science Ignores Africa's Native Crops. *afrol News*. Available at: http://www.afrol.com/articles/22332. Accessed Novermber 21, 2006.

34. Johns T. Plant biodiversity and malnutrition: Simple solutions to

complex problems. *Afr J Food Agric Nutr Dev.* 2003; 3:45-52.

35. Saran S, Gopalan S, Krishna T. Use of fermented foods to combat stunting and failure to thrive. *Nutrition.* May 2002; 18(5):393-396.

36. Pavan L, Casiglia E, Pauletto P, et al. Blood pressure, serum cholesterol and nutritional state in Tanzania and in the Amazon: Comparison with an Italian population. *J Hypertens.* October 1997; 15(10):1083-1090.

37. Pauletto P, Puato M, Caroli M, et al. Blood pressure and atherogenic lipoprotein profiles of fish-diet and vegetarian villagers in Tanzania: the Lugalawa study. *Lancet.* Sept 1996; 348(9030):784-788.

38. Marles R, Farnsworth N. Anti-diabetic plants and their active constituents. *Phytomedicine.* 1995; 2:137-189.

39. Johns T, Sthapit B. Biocultural diversity in the sustainability of developing-country food systems. *Food Nutr Bull.* June 2004; 25(2):143-155.

40. Chweya J, Eyzaguirre P. *The Biodiversity of Traditional Leafy Vegetables.* Rome: International Plant Genetics Resource Institute (IPGRI); 1999.

41. Zagre N, Delpeuch F, Traissac P, Delisle H. Red palm oil as a source of vitamin A for mothers and children: Impact of a pilot project in Burkina Faso. *Public Health Nutr.* Dec 2003; 6(8):733-742.

42. Uiso F, Johns T. Consumption patterns and nutritional contribution of Crotalaria brevidens (mitoo) in Tarime District, Tanzania. *Ecol Food Nutr.* 1996; 35(1):59-69.

43. Fassil H, Guarino L, Sharrock S, Bhagmal S, Hodgkin T, Iwanaga M. Diversity for food security: Improving human nutrition through better evaluation, management, and use of plant genetic resources. *Food Nutr Bul.* 2000; 21:497-502.

44. West C, Temalilwa C. *The Composition of Foods Commonly Eaten in East Africa*. Wageningen: Wageningen Agricultural University; 1988.
45. Johns T. Dietary diversity, global change, and human health. Paper presented at: Proceedings of the symposium " Managing Biodiversity in Agricultural Ecosystems". 8-10 November, 2001; Montreal, Canada.
46. Fahey J. Underexploited African grain crops: A nutritional resource. *Nutr. Rev.* 1998; 56(9):282-258.
47. Gibson RS. Traditional methods for food processing diet modification and diversity to increase micronutrient availability. *Ann Nutr Metab.* 2005; 49(Suppl. 1):10.
48. Johns T, Faubert G, Kokwaro J, Mahunnah R, Kimanani E. Anti-giardial activity of gastrointestinal remedies of the Luo of East Africa. *J Ethnopharmacol.* April 1995; 46(1):1-23.
49. Johns T, Nagarajan M, Parkipuny M, Jones P. Maasai gummivory: Implications for paleolithic diets and contemporary health. *Curr Anthropol.* 2000; 41:453-459.
50. Lindhorst K. *Antioxidant Activity of Phenolic Fraction of Plant Products Ingested by the Maasai.* [M.Sc. thesis]. Montreal, Canada, McGill University; 1998.
51. Chapman L, Johns T, Mahunnah R. Saponin-like *in vitro* characteristics of extracts from selected non-nutrient wild plant food additives used by Maasai in milk and meat based soups. *Ecol Food Nutr.* 1997; 36:1-22.
52. Johns T. The chemical ecology of human ingestive behaviors. *Annu Rev Anthropol.* 1999; 28:27-50.
53. Johns T, Eyzaguirre P. Nutrition for sustainable environments. *SCN News.* 2000; 21:24-29.

54. Messer E. *Mehtods for Studying Determinants of Food Intake*. In: Pelto G, Pelto P, Messer E, eds. Research Methods in Nutritional Anthropology. Hong Kong: The United Nations University; 1989:1-201.
55. Frison E, Cherfas J, Eyzaguirre P, Johns T. Biodiversity, nutrition and health: making a difference to hunger and conservation in the developing world. Paper presented at: Keynote Address to the Seventh Metting of the Conference of the Parties to the Confention on Biological diversity (COP7). 2004.
56. IndigenoVeg. Networking to Promote the Sustainable Production and Marketing of Indigenous Vegetables through Urban and Peri-Urban Agriculture in Sub-Saharan Africa (IndigenoVeg). Available at: http://www.indigenoveg.org/. Accessed 26.03.2006.
57. Johns T, Kimiywe J, Waudo J, Mutemi E, Maundu P. Traditional dietary diversity against the nutrition transition: An East African case study. *Ann Nutr Metab.* 2005; 18(Suppl. 1):36.
58. FANUS. The federation of African nutrition societies - online. Available at: http://www.africanutrition.org/. Accessed November 27, 2006.
59. Hoffmeister M, Lyaruu I, Krawinkel M. Assessment of nutritional intake, body mass index and glycemic control in patients with type-2 diabetes from northern Tanzania. *Ann Nutr Metab.* 2005; 49(1):64-68.
60. Smith I. *Foods of West Africa:* Their Origin and Use. Ottawa, Canada; 1998.
61. Gura S. A note on traditional food plants in East Africa: Their value for nutrition and agriculture. *Food Nutr.* 1986; 12(1):18-22.

DISCUSSION

Recent non-communicable disease (NCD) epidemics (i.e. diabetes, hypertension, obesity, and cardiovascular disease) in sub-Saharan Africa appear to be the result of a *nutrition transition* whereby traditional African food habits have been progressively replaced by a *globalized food culture*.[1,2] Moreover, non-communicable, chronic diseases have not simply replaced infectious and malnutrition related disease in this region. These vulnerable populations now experience a *double burden* of disease, where the effects of the *nutrition transition* are added to the existing problem.[3,4]

As the people of this region continue to experience a *nutrition transition*, ancient indigenous knowledge about traditional African foods and food habits will continue to wane.[5,6] Nutrition is coming to the forefront as a major modifiable determinant of NCDs, with scientific evidence increasingly supporting the view that long-term improvement of food habits can have a profound, positive effect on health status throughout the lifetime. The public health approach of primary prevention, which includes a locally diverse and traditional diet, is considered to be the most cost-effective, affordable and sustainable course of action to reduce the upward trend of NCDs.[7] The primary intent of this project was to investigate the health-related importance of traditional East African food habits and advance awareness of the richness of the African food culture. The investigation culminated in the development of an innovative online collection (web site) (which was available at: http://www.healthyeatingclub.org/Africa/ and has been transferred to http://www.drverena.com).[8]

The online collection was primarily based upon a unique and precious set of studies obtained through the activities of the *Max-Planck*

Nutrition Research Unit, formerly located in Bumbuli, Tanzania. These studies, conducted from the 1930s to 1960s, provide the first empirical evidence of traditional food habits collected from East Africa, including the countries of Tanzania, Zanzibar Island, Pemba Island, Kenya, and Uganda.[9] Professor Ulrich Oltersdorf, who was also involved in some of the research at the *Max-Planck Nutrition Research Unit*, graciously made the studies available for the purpose of contributing significantly to the development of an online collection. The series of studies has therefore been entitled the *Oltersdorf Collection*.[9] At present, the online information system[8] serves as an important research and educational tool for nutritional scientists and the general public interested in investigating traditional African food habits.

The investigations of Chapter 1, based on: (1) a systematic online search, performed to assess current gaps in online collections, and (2) a questionnaire, administered to opinion leaders in the nutritional sciences at the 18th International Congress on Nutrition (ICN) in Durban, South Africa, in September 2005, revealed several important findings.[10] According to the systematic search, there are currently no online collections that have an overall focus on traditional African food habits. Moreover, 82% of the opinion leaders, interviewed at the recent INC believed that a gap currently exists in the online dissemination of such information. The findings of the survey also supported the position that Africa continues to experience a *nutrition transition*. According to the interviewees, the adoption of western values, urbanization, economic pressures, maldistribution of wealth, and scarcity through lack of choice were primary factors driving the *nutrition transition* and the related *double burden* epidemic in Africa today. The increasing prevalence of NCDs associated with these socio-economic pressures has been well described.[1, 11-14] Further, the experts surveyed (n=92) agreed that traditional African foods and food habits were superior to the globalized food habits currently underpinning the *nutrition transition*. The respondents believed that the knowledge of traditional African food habits is being lost (80%) and that there is a vital need for investigation and documentation to preserve this knowledge.

Based on this strong rationale, an online collection (web site) (available at: www. healthyeatingclub.com/Africa/), aimed at preserving knowledge of traditional foods and food habits in Africa was developed (See Chapter 2). With the primary intent of collating data for this online information system, the *Oltersdorf Collection* was reviewed. The data extraction revealed that a potential did exist for a rich food culture in East Africa from the 1930s to the 1960s, despite years of empirical occupation. Many of the traditional foods that have been presented in Chapter 2 have known health benefits according to the latest annals of scientific inquiry. A diversified diet combined with traditional knowledge of food preparation may advance our understanding of health by complementing current scientific knowledge of micronutrient density and nutrient absorption. The knowledge which has evolved and remains at the heart of this *Cradle of Civilization*, including incredible knowledge of the relationship between food, food habits and longevity, should not be ignored and should indeed be investigated further. Such information can and should be utilized by the global community for improving the current *globalized* food culture, which has largely been responsible for the obesity and diabetes epidemics currently plaguing the world.

To support the contention that traditional African food habits are associated with significant health benefits, a systematic review of the *Oltersdorf Collection* was conducted (See Chapter 3), to investigate relationships between dietary intake and health status indices investigated within specific cohorts in East Africa (i.e. Tanzania, Kenya and Uganda) from the 1930s to the 1960s. Overall, the review revealed that many ethnic groups did not exhibit adequate dietary intake and did not consume a diversity of traditional whole foods representative of the wide spectrum of food choices available in the region at this time. While NCDs were not prevalent, there was substantial reporting of malnutrition-related and infectious diseases, particularly among children. The review presented in Chapter 3 suggests that the shift from a traditional, diversified diet to a simplified, monotonous diet may have been concomitant with the onset of cash-crop farming in East Africa. This particular finding led to the investigation of additional factors that may have been implicated in the dietary simplification of East Africans during this period.

Chapter 4 revealed that numerous factors have underpinned the *nutrition transition* in Kenya, Uganda and Tanzania, from early colonization to the current oppressive, political-economic structure. It is imperative that greater efforts be directed toward exposing these forces, and proposing solutions to the *nutrition transition* in Africa. Without thorough investigation, documentation and widespread dissemination of this information, efforts to improve the NCD epidemics will prove futile and the vast continent of Africa and its people will continue to suffer from epidemics of chronic diseases.

Although the *nutrition transition* has affected much of East Africa today, there remain some cohorts which still consume a traditional diet. Chapter 5 was undertaken to determine if adherence to a traditional East African diet has been associated with better markers of health status, including a lower NCD risk factor profile, versus adherence to a non-traditional diet. The studies included in the review provided limited information regarding the intake of micro and macronutrients and the composition of meals in the cohorts studied, making the data difficult to interpret. However, the studies reviewed provide some support for the health related benefits of the traditional East African diet versus a non-traditional diet, particularly with regard to NCD risk factors such as hypertension, dyslipidaemia and obesity. The studies reviewed also provide some support for the protective effects of increased fish consumption, particularly on blood lipid profiles (i.e. dyslipidemia). Additional research is likely required to more thoroughly document traditional diets amongst the East African population, and investigate relationships between dietary intake and health status indices. Such research is needed to identify the magnitude and impact of the *nutrition transition* on food habits and the prevalence of NCDs in East Africa.

To preserve and disseminate knowledge of traditional African foods and food habits, an online collection was created which has been available since May 2006, via the worldwide web.[8] The purpose of Chapter 6 was to determine the overall utilization of the online collection over these initial 31 weeks, and the discussion of its potential future applications. Overall, visits to the online collection increased by 17%, from week 1 to week 31. On average, the web site accounted for 2958 visits per week. These results suggest that the online collection is increasing in popularity, and is frequently accessed for various topics and PDF files.

It can be hypothesized that our online collection can successfully contribute to the documentation, compilation, and dissemination of information pertaining to traditional African food habits. The future development of the web site project will include the amalgamation of research, both current and historical, pertaining to traditional foods and food habits throughout all regions of Africa. In addition, the web site may enhance the development of an online network of communication (i.e. a research forum), both within Africa and abroad, for the development of targeted and relevant collaborative research projects.

References

1. Maletnlema T. A Tanzanian perspective on the nutrition transition and its implication for health. *Public Health Nutr.* 2002; 5(1A):163-168.
2. Popkin B, Lu B, Zhai F. Understanding the nutrition transition: Measuring rapid dietary changes in transitional countries. *Public Health Nutr.* 2002; 5(6A):947-993.
3. Popkin B. An overview of the nutrition transition and its health implications: The Bellagio meeting. *Public Health Nutr.* 2002; 5:93-103.
4. Popkin B. The nutrition transition and prevention of diet related disease in Asia and the Pacific. *Food Nutr Bull.* 2001; 22:S1-58.
5. Oniang'o RK. Food habits in Kenya: The effects of change and attendant methodological problems. *Appetite.* 1999; 32:93-96.
6. Kuhnlein HV, Johns T. Northwest African and Middle Eastern food and dietary change of indigenous peoples. *Asia Pac J Clin Nutr.* 2003; 12(3):344-349.
7. World Health Organization. *Diet, nutrition and the prevention of chronic diseases. Report of a Joint WHO/FAO Expert Consultation.* Geneva 2003.
8. Raschke V. East African food habits on-line. In: Wahlqvist ML. *HEC Press* Available at: http://www.healthyeatingclub.org/Africa/index.htm.
9. Raschke V, Oltersdorf U, Elmadfa I, Wahlqvist M, Cheema B, Kouris-Blazos A. Content of a novel online collection of traditional East

African food habits (1930s - 1960s): Data collected by the *Max-Planck Nutrition Research Unit*, Bumbuli, Tanzania. *Asia Pac J Clin Nutr.* 2007; 16(1):140-151.

10. Raschke V, Oltersdorf U, Elmadfa I, Wahlqvist M, Kouris-Blazos A, Cheema B. The need for an online collection of traditional African food habits. *Afr J Food Agr Nutr Dev.* Submitted May 2006; In Press.

11. Drewnowski A, Popkin B. The nutrition transition: New trends in the global diet. *Nutr Rev.* 1997; 55(2):31-43.

12. Levitt N, Katzenellenbogen J, Bradshaw D, Hoffman M, Bonnici F. The prevalence and identification of risk factors for NIDDM in urban Africans in Cape Town, South Africa. *Diabetes Care.* 1993; 16(4):601-607.

13. Bourne LT, Langenhoven ML, Steyn Kea. Nutritional intake of the African population of the Cape Peninsula, South Africa: The BRISK study. *Cent Afr J Med.* 1993; 39:238-247.

14. Trowell H. From normotension to hypertension in Kenyans and Ugandans 1928-1978. *East Afr Med J.* March 1980; 57(3):167-173.

I want morebooks!

Buy your books fast and straightforward online - at one of the world's fastest growing online book stores! Environmentally sound due to Print-on-Demand technologies.

Buy your books online at
www.get-morebooks.com

Kaufen Sie Ihre Bücher schnell und unkompliziert online – auf einer der am schnellsten wachsenden Buchhandelsplattformen weltweit!
Dank Print-On-Demand umwelt- und ressourcenschonend produziert.

Bücher schneller online kaufen
www.morebooks.de

OmniScriptum Marketing DEU GmbH
Heinrich-Böcking-Str. 6-8
D - 66121 Saarbrücken
Telefax: +49 681 93 81 567-9

info@omniscriptum.com
www.omniscriptum.com

Printed by Books on Demand GmbH, Norderstedt / Germany